Dealing with Disaster in Japan

Just as the sinking of the *Titanic* is embedded in the public consciousness in the English-speaking world, the crash of JAL flight JL123 is part of the Japanese collective memory. The 1985 crash involved the largest loss of life for any single air crash in the world. Five hundred and twenty people, many of whom had been returning to their ancestral home for the Obon religious festival, were killed; there were only four survivors.

This book tells the story of the crash, discusses the many controversial issues surrounding it and considers why it has come to have such importance for many Japanese. It shows how the Japanese responded to the disaster: trying to comprehend how a faulty repair may have caused the crash, and the fact that rescue services took such a long time to reach the remote crash site; how the bereaved dealt with their loss; how the media in Japan and in the wider world reported the disaster; and how the disaster is remembered and commemorated. The book highlights the media coverage of anniversary events and the Japanese books and films about the crash, the very particular memorialization process in Japan, alongside Japanese attitudes to death and religion, and it points out in what ways this crash both reflects typical Japanese behaviour and in what ways the crash is unique.

Christopher P. Hood is a Reader in Japanese Studies at Cardiff University, UK. His publications include: *Shinkansen: From Bullet Train to Symbol of Modern Japan* (2006), *Japanese Education Reform: Nakasone's Legacy* (2001) and (as editor) the four volume *Politics of Modern Japan* (all published by Routledge).

Routledge Contemporary Japan Series

1 A Japanese Company in Crisis
Ideology, strategy, and narrative
Fiona Graham

2 Japan's Foreign Aid
Old continuities and new directions
Edited by David Arase

3 Japanese Apologies for World War II
A rhetorical study
Jane W. Yamazaki

4 Linguistic Stereotyping and Minority Groups in Japan
Nanette Gottlieb

5 Shinkansen
From bullet train to symbol of modern Japan
Christopher P. Hood

6 Small Firms and Innovation Policy in Japan
Edited by Cornelia Storz

7 Cities, Autonomy and Decentralization in Japan
Edited by Carola Hein and Philippe Pelletier

8 The Changing Japanese Family
Edited by Marcus Rebick and Ayumi Takenaka

9 Adoption in Japan
Comparing policies for children in need
Peter Hayes and Toshie Habu

10 The Ethics of Aesthetics in Japanese Cinema and Literature
Polygraphic desire
Nina Cornyetz

11 Institutional and Technological Change in Japan's Economy
Past and present
Edited by Janet Hunter and Cornelia Storz

12 Political Reform in Japan
Leadership looming large
Alisa Gaunder

13 Civil Society and the Internet in Japan
Isa Ducke

14 Japan's Contested War Memories
The 'memory rifts' in historical consciousness of World War II
Philip A. Seaton

15 Japanese Love Hotels
A cultural history
Sarah Chaplin

16 Population Decline and Ageing in Japan
The social consequences
Florian Coulmas

17 Zainichi Korean Identity and Ethnicity
David Chapman

18 A Japanese Joint Venture in the Pacific
Foreign bodies in tinned tuna
Kate Barclay

19 Japanese–Russian Relations, 1907–2007
Joseph P. Ferguson

20 War Memory, Nationalism and Education in Post-War Japan, 1945–2007
The Japanese history textbook controversy and Ienaga Saburo's court challenges
Yoshiko Nozaki

21 A New Japan for the Twenty-First Century
An inside overview of current fundamental changes and problems
Edited by Rien T. Segers

22 A Life Adrift
Soeda Azembo, popular song and modern mass culture in Japan
Translated by Michael Lewis

23 The Novels of Oe Kenzaburo
Yasuko Claremont

24 Perversion in Modern Japan
Psychoanalysis, literature, culture
Edited by Nina Cornyetz and J. Keith Vincent

25 Homosexuality and Manliness in Postwar Japan
Jonathan D. Mackintosh

26 Marriage in Contemporary Japan
Yoko Tokuhiro

27 Japanese Aid and the Construction of Global Development
Inescapable solutions
Edited by David Leheny and Carol Warren

28 The Rise of Japanese NGOs
Activism from above
Kim D. Reimann

29 Postwar History Education in Japan and the Germanys
Guilty lessons
Julian Dierkes

30 Japan-Bashing
Anti-Japanism since the 1980s
Narelle Morris

31 Legacies of the Asia-Pacific War
The Yakeato generation
Edited by Roman Rosenbaum and Yasuko Claremont

32 Challenges of Human Resource Management in Japan
Edited by Ralf Bebenroth and Toshihiro Kanai

33 Translation in Modern Japan
Edited by Indra Levy

34 Language Life in Japan
Transformations and prospects
Edited by Patrick Heinrich and Christian Galan

35 The Quest for Japan's New Constitution
An analysis of visions and constitutional reform proposals 1980–2009
Christian G. Winkler

36 Japan in the Age of Globalization
Edited by Carin Holroyd and Ken Coates

37 Social Networks and Japanese Democracy
The beneficial impact of interpersonal communication in East Asia
Ken'ichi Ikeda and Sean Richey

38 Dealing with Disaster in Japan
Responses to the Flight JL123 crash
Christopher P. Hood

Dealing with Disaster in Japan

Responses to the flight JL123 crash

Christopher P. Hood

Routledge
Taylor & Francis Group
LONDON AND NEW YORK

This edition published 2012
by Routledge
2 Park Square, Milton Park, Abingdon, Oxon OX14 4RN

Simultaneously published in the USA and Canada
by Routledge
711 Third Avenue, New York, NY 10017

Routledge is an imprint of the Taylor & Francis Group, an informa business

British Library Cataloguing in Publication Data
A catalogue record for this book is available from the British Library

Library of Congress Cataloging-in-Publication Data
Hood, Christopher P. (Christopher Philip), 1971-
Dealing with disaster in Japan : responses to the flight JL123 crash / Christopher P. Hood.
p. cm. -- (Routledge contemporary Japan series ; 38)
Includes bibliographical references and index.
1. Aircraft accidents--Japan--1985. 2. Aircraft accident victims--Japan. 3. Bereavement--Japan. 4. Japan--Social life and customs. 5. Nihon Koku. I. Title.
TL553.53.J3H66 2011
363.12'480952--dc22
2011009024

ISBN: 978-0-415-45662-3 (hbk)
ISBN: 978-0-203-80450-6 (ebk)

Typeset in Times New Roman by
GreenGate Publishing Services, Tonbridge, Kent

MIX
Paper from responsible sources
FSC® C004839
www.fsc.org

Printed and bound in Great Britain by
CPI Antony Rowe, Chippenham, Wiltshire

To the passengers and crew of JL123, their families and those who helped them,

and

to Monty.

Contents

Figures

Photographs taken by the author can be seen at www.hood-online.co.uk/JL123.

Tables

Notes on style

The revised Hepburn Romanization system is used for Japanese words. Consequently, the prefecture where JL123 crashed is written as Gunma rather than Gumma. An apostrophe is used between combinations 'n' and 'yo', for example, to distinguish between the sounds *n'yo* and *nyo*. Macrons are used as appropriate on Japanese words. Macrons denote long vowel sounds and are twice the length of a short vowel in pronunciation. Macrons have been added as appropriate to words in quotations where they were omitted in the original. Macrons are not included on words where the brand name does not include them, e.g. Tokyo Disneyland.

All personal names are written in the order: given name, family name.

Place names are given in their proper local spelling, e.g. Napoli rather than Naples.

Dates, when abbreviated, are written according to the ISO 8601 international standard (which is similar to the Japanese system). For example, 12 August 1985 becomes 1985-08-12.

References (to show sources of information) are indicated by a superscript number. The reference itself uses the Harvard style and is presented in the References section after the main text. To distinguish notes from references, superscript letters are used and the note appears in the Notes section after the main text.

In 2001 the Ministry of Transport merged with some other ministries to become the Ministry of Land, Infrastructure and Transport (MLIT, *Kokudokōtsūshō*). Its English name subsequently became the Ministry of Land, Infrastructure, Transport and Tourism in 2008, although its acronym and Japanese name remained unchanged. For simplicity, the term Ministry of Transport is used throughout this book.

There are numerous airports in Japan and their names do not always relate to their exact location. From its opening in 1978 until 2004, Narita International Airport (NRT) was known as New Tōkyō International Airport, despite being located in Chiba Prefecture and some 60km from Tōkyō. In 1978 Tōkyō International Airport (HND), located on a man-made island in Tōkyō Bay, became largely a domestic airport until the expansion of the airport in 2010. It is usually known as Haneda Airport. Ōsaka International Airport (ITM), is essentially a domestic airport and is often referred to as Itami Airport. Much of the airport is located in Itami city, which is in Hyōgo Prefecture, although the terminal building and main access is in Toyonaka city (Ōsaka Prefecture). Kansai International Airport (KIX) opened in 1994 and is now the main international airport for Ōsaka. It is often referred to as Kankū in Japanese. Sapporo's main airport, located in Chitose and Tomakomai cities about 30 km from Sapporo, is New Chitose Airport (CTS), which opened in 1988 to replace Chitose Airport, which subsequently became an airbase for the Japanese Air Self-Defense Forces. The current airport tends to be referred to as Chitose Airport. Throughout the text these airports are referred to as Narita, Haneda, Itami, Kansai and Chitose.

Japan Air Lines changed its English name to Japan Airlines in 1989. To avoid confusion I use the acronym JAL, which the company has used throughout, and is the company's ICAO (International Civil Aviation Organization) airline designator. Throughout the text I use 'JL123' to refer to the flight number. Although it is commonly referred to as 'JAL 123', flight numbers are usually referred to using a company's IATA (International Airline Transport Association) airline designator, e.g. JL for JAL, NH for All Nippon Airways, etc. It should be noted that JAL tend to refer to the crash by the plane's unique registration number, JA8119.

Throughout the text the crash site is referred to as Osutaka-no-One rather than Mount Osutaka or Osutaka-yama. Hyphens have been added to avoid confusion with the English words 'no' and 'one' (note that the letter 'e' in pronounced like a French 'é'). The village in which the plane crashed is referred to as Ueno-mura rather than Ueno village or Ueno (which is a common place name).

Where translations have not been attributed to other sources, they are my own.

British English spelling is used except when citing a proper name which uses American English spelling. Citations have been altered to be consistent with this where necessary. The universal numbering system (adopted officially in 1974) is used, i.e., 1 billion = 1,000,000,000 and 1 trillion = 1,000,000,000,000.

How some of the conventions were decided upon is discussed in the main text.

Metric (SI) measurements are generally used throughout the text. However, feet, miles and nautical miles are still regularly used in the aviation world. Where appropriate the original measurement has been used (see Glossary for details about conversions). Nautical miles (NM) and knots are non-SI units used within the SI system.

Due to fluctuations in exchange rates, conversions are not used. On 12 August 1985, £1 = ¥328 and $1 = ¥237. On 12 August 2010, £1 = ¥133 and $1 = ¥86.

Preface and acknowledgements

Twenty-five years ago in the area where I am writing this preface, flight JL123 crashed. Today it is the focal point for remembering those who died on JL123. As well as *izoku*, one can see those who have supported them over the years. There are also many reporters and photographers. What happened here, how they are remembered, what the *izoku* have done, who supported them, and why there continues to be so much interest in the crash has been the focus of my research for the past three and a half years.

This research would not have been possible without the help of many people and organizations. I would first like to thank Peter Mathews and Keith Haines, for it was communications with them that helped me to decide that I should do this study. Peter tragically lost his son in the crash, and Keith was the JAL employee who was assigned to look after the Mathews family. Peter very generously gave me full access to the diary he kept in 1985 as well as the photographs he took. The information contained within them enabled me to get a much clearer understanding of what it was like for some *izoku* in 1985. Keith's comments ensured that I got a clearer understanding of what it was like for those JAL employees who were assigned to looking after *izoku*.

When I began this research I did not expect to speak to many, if any, *izoku* other than Peter, but I found that *izoku* wanted to speak to me – to find out about why I was in Ueno-mura and to tell me about their experiences. I would like to thank all those who have provided information in this way. I would particularly like to thank Susanne Bayly, Mariko Kawaguchi, Shinobu Kawahara, Takao Kawakita, Corinna Klaubert, Shigeko Masunaga, Kuniko Miyajima, Ryōichi Ogawa, Kimi and Hideo Ozawa, Hitomi Taoka and Diana Yukawa.

I would also like to thank the various other people I interviewed for my research and those who provided other information and help. In particular I would like to thank from JAL: Akemi Hamazaki, Yutaka Kanasaki, Kiyoshi Kishida, Shinobu Kobayashi, Stephen Pearlman, Tadao Sakai, Kōjirō Waki and Tamotsu Watanabe; from *Jōmō Shimbun*: Mitomu Arai, Hiroshi Fujii and Motofumi Kabashima; from *Kyōdo News*: Yasushi Fujii, Will Hollingsworth and Miya Tanaka; from the *Mainichi Shimbun*: Shinji Hagio; from the Ministry of Defense: Fumito Matsushima and Kentarō Sakagawa; from *Sankei Shimbun*: Shintarō Harakawa, Katsuji Magishi and Tomohiko Uchiyama; from Shinohara Dental Clinic: Mizuo

Shinohara and Eiko Kosuge; from the *Yomiuri Shimbun*: Miyuki Konnai. I would also like to thank Glen Fukushima (Airbus Japan), Mike Ghiglione, Gill Goddard (Sheffield University library), William Horsley (BBC correspondent in Tōkyō in 1985), Masaaki Ikeda (author), Kentarō Katō (Fuji TV), Tadakazu Kimura (*Asahi Shimbun*); Colonel Yasuyuki Kodama (Ground Self-Defense Forces), Naonori Kohira (photographer), Rikei Kubo (producer of *Kuraimāzu Hai* (2008)); Detective Chief Inspector Jan MacKay (Public Protection Department, South Wales Police), Masaaki Osawa (Governor, Gunma Prefecture), Fumuhiko Ōshima (Fire and Disaster Management Agency), Lawrence Repeta, Prof. Ian Reader (Manchester University), Kōichi Takamoto (ALPA Japan), Frank Taylor (aircraft crash investigator), Akira Teramura (formerly at Tōkyō Fire Department), Geoffrey Tudor (formerly at JAL now at Kenyon Worldwide Disaster Management), Hisaaki Wakaizumi (NHK, producer of *Kuraimāzu Hai* (2005)), Andrew Weir (author), Akio Yamazaki (author), Keiichi Yamazaki (amateur photographer), Kunio Yanagida (author), Shinobu Yoshioka (author), Hideo Yokoyama (author) and Shinsaku Zama (National Research Institute of Fire and Disaster).

I would also like to thank a number of organizations which have provided help and information. In particular I would like to thank Fuji-Sankei Communications International, JAL, *Jōmō Shimbun*, *Kyōdo News*, NAVIX and Video Research Ltd. Thanks also to those individuals and companies, named throughout the text, who gave me permission to use photographs and images, especially those kind enough not to charge any fee or only a nominal amount. A word of thanks also to Routledge for putting up with my various demands and particularly Peter Sowden whose support has always been invaluable.

Parts of my research have been presented at a variety of conferences over the years and I would like to thank all those who attended my papers and provided insights and asked questions which helped me to further develop my research. I would also like to give a special mention to my final year students in 2009–10 who studied *Kuraimāzu Hai* as part of one of their modules, and whose questions and ideas about that story, its dramatizations and my research also helped me to crystallize some of my ideas. I would also like to thank those who filled in my online survey and those YouTube users who responded to my enquiries.

I would also like to thank the various companies and organizations that have helped me pursue this research through the provision of funding: Cardiff Business School, Daiwa Anglo-Japanese Foundation, Great Britain Sasakawa Foundation, the Japan Foundation Endowment Committee and the Japan Foundation (London Office).

I would like to thank all my family and friends who helped to keep me going. In particular I would like to thank the Gotō family, Suguru Funatsu, Peter Matanle, Rie Ōkubo, Takako Pocklington, the Tanabe family and the Tanaka family. Finally, a special word of thanks to Man Yee, Miella and Monty – you all fill up my senses.

Christopher Hood
Osutaka-no-One
12 August 2010

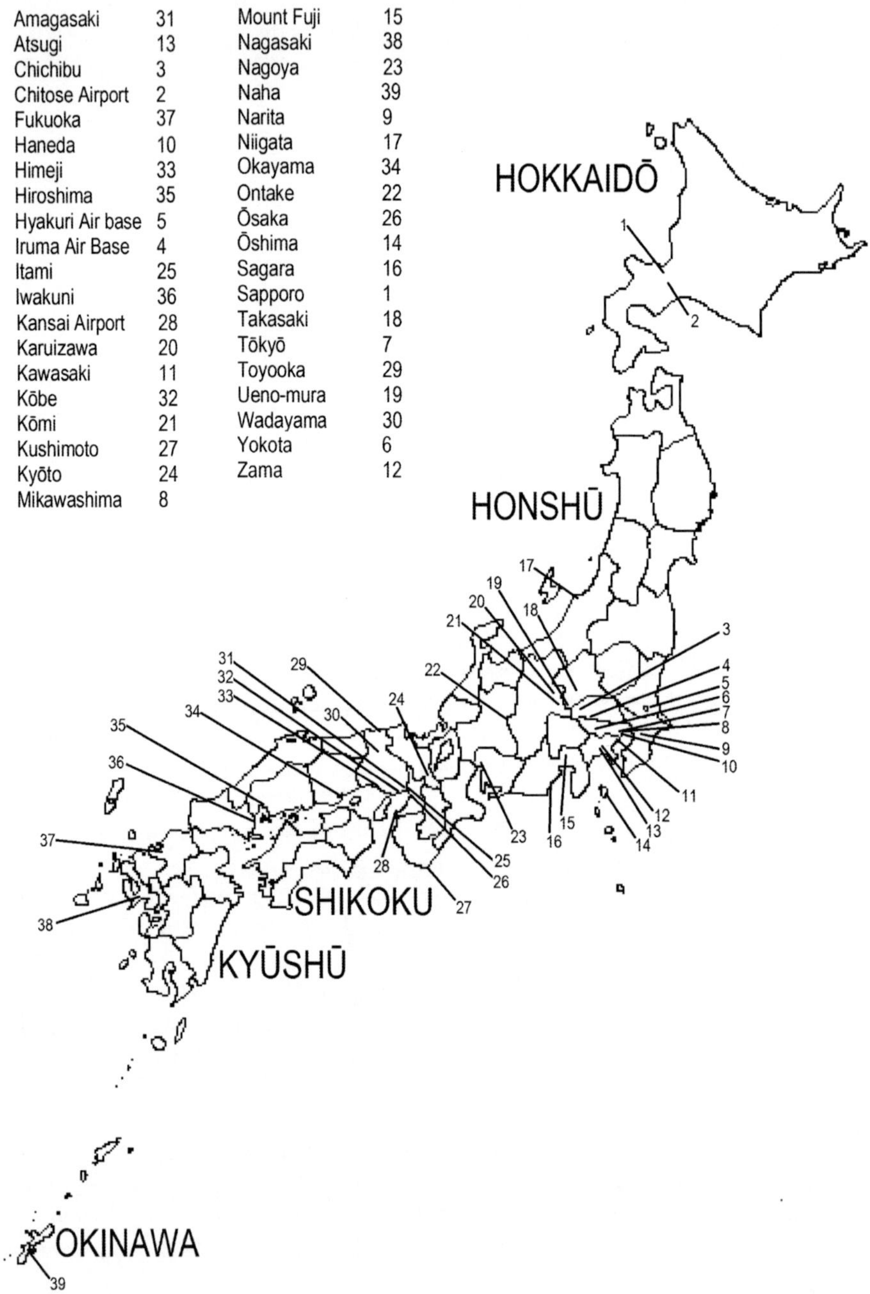

Figure 0.1 Map of Japan

Map by the author.

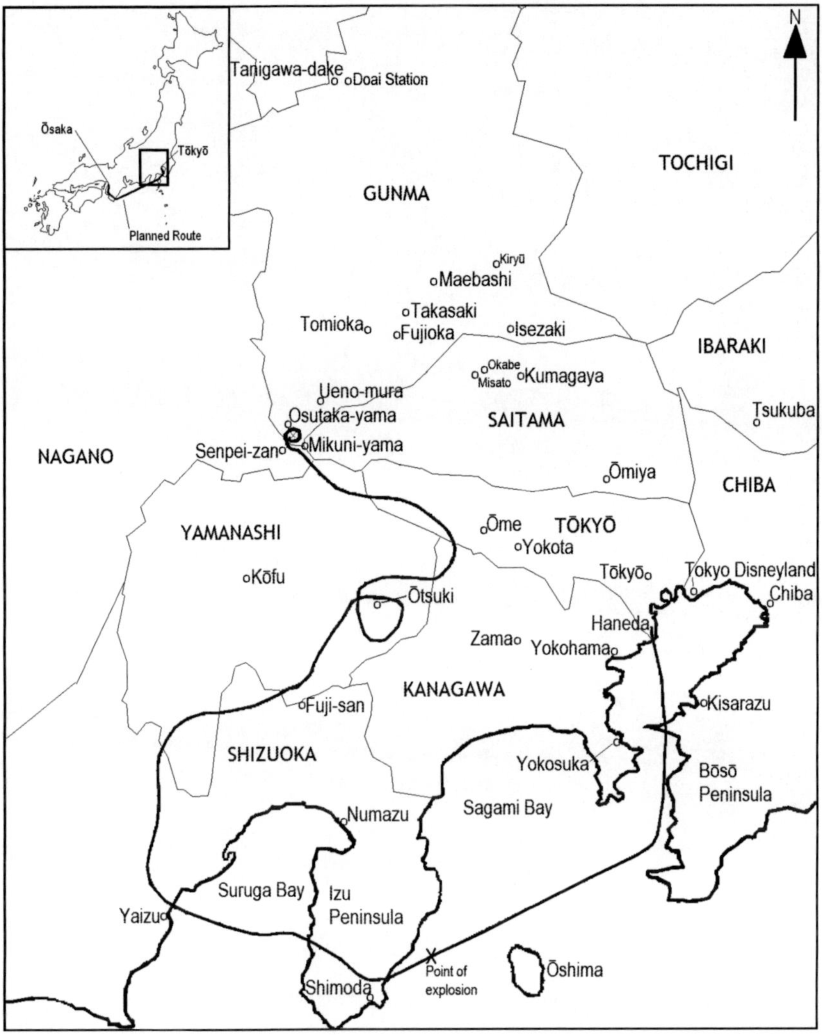

Figure 0.2 JL123 route map

Map by the author based on map in AAIC 1987b: Figure 1.

Part I
Context

1 Introduction

Do not stand at my grave and weep,
I am not there;
I do not sleep.
I am a thousand winds,
A thousand winds that blow in the open sky.[a]

One day you will die. That we die is one of the few certainties about our lives. Death connects us all. How we will die is a thought, and sometimes a fear, that we may ponder throughout our lives. Many are fortunate to pass away quietly. For others the way in which they die will be more dramatic. Sometimes these events can make an impression upon many more lives than just the relations and friends of those killed. This study is of one such event.

Flight JL123 crashed with the loss of 520 lives. Then, and at the time of writing, it is the largest number of deaths from a single plane crash. But why is there still so much interest in this crash, not only in Japan, but around the world? This is one of the questions that this book looks to answer and was one of the starting points for the research itself. Although the book considers the cause of the crash, this study is much more concerned with the human side of the tragedy and what can be learnt about Japanese society from looking at the responses to it. It is for that reason that the names of the crew and passengers are listed in the footer of pages of the book. This study is not about a lifeless mixture of metal and other man-made objects falling out of the sky, but about real people. Whenever you read 'JL123', please look to the footer of the page to remind yourself of this important fact. The survivors' names are also included in the list and are written in bold. Other issues related to presenting the names are discussed elsewhere in the book.

Some of the text is quite graphic and probably disturbing to some readers. I hope so. What is written represents only a fraction of some of the details which are provided in a number of Japanese sources that I have read, as will be discussed further in Chapter 8. This was a human tragedy. Real people died in the crash. Real people suffered then and many continue to suffer in various forms today. This must not be forgotten. Perhaps if more people truly understood the nature of what happens following an aircraft accident there would be greater pressure

Masami Takahama 高浜雅己 *(49)* · *Yutaka Sasaki* 佐々木祐 *(39)* · *Hiroshi Fukuda* 福田博 *(46)*

to make air travel even safer and to improve the way in which *izoku* and others connected to a crash are subsequently cared for. I had concerns about areas of this book being too graphic and insensitive to *izoku,* but while such detail is unusual in English-language books, it is not in Japanese ones, so I assume my concerns are misplaced. Additionally, whatever I write must be far less graphic than what many had to see for themselves, so I hope that all readers can understand why the text is presented in this way.

The book is divided into five parts and nine chapters. Each part has its own specific role and its own questions to answer. Part I is concerned with putting the study into context. This chapter starts off by providing some answers to the questions I have frequently been asked, such as why I choose to do this research. It then goes on to explain how the research was done and some of the challenges that I faced. Finally, this chapter explains how some of the conventions set out in the 'Notes on style' section were decided upon and raises broader questions about how those in Japanese studies present their research. Chapter 2 is also primarily concerned with putting the study into context. Although contextual information will appear throughout the book, one of the concerns I had was that readers may not have an appreciation of some of the generic issues; what was the world like in 1985? Why was a jumbo jet being used on a domestic flight? What was the history of JAL up to 1985? What other disasters – 'natural' and 'man-made' – have there been in Japan?

The main body of the book is split into three parts, and at the start of each part the timeline resets back to 12 August 1985. Part II is concerned with the crash itself and the investigation into its cause. When I started this research I had no plans to discuss the cause of the crash; I am not an expert in aviation engineering or crash investigation, and, based on my knowledge of the crash, there appeared no reason to do any investigation. However, as I started reading Japanese books I discovered that there were a number of concerns being raised about the conclusions of the official investigation. I believe that, as this is the first book in English to be solely about JL123, there is a need to present some of what has been discussed in Japanese sources. Indeed, the difference in views about the cause of the crash between the English-speaking world and Japanese-speaking world and how to address this became one of the concerns of my research. Chapter 3 deals with the events surrounding the crash itself. I have deliberately kept this chapter in a narrative form and have avoided including a discussion of the cause. This discussion is provided in the subsequent chapters. Chapter 4 deals with the response to the crash and with issues of disaster management in Japan. Why did it take so long to locate the crash? Why did it take so long for rescue teams to get to the crash site? How was the cause of the crash investigated and what was the conclusion? Who was responsible for the crash?

Part III looks at the human cost of the crash. Chapter 5 covers what happened in the hours, days and weeks after the crash as bodies and remains had to be identified and the families sought. How were the remains identified? Why did some passengers write *isho*? What can we learn from the photographs taken by one of the passengers during the final 32 minutes of the flight? How were the *izoku* cared

Jun Hatano 波多野純 *(39)* · *Mariko Akata* 赤田真理子 *(31)* · *Mitsuyo Ebina* 海老名光代 *(28)*

for? How were the *izoku* compensated for their loss? Chapter 6 looks at the longer term response of the families and how they remember their loved ones. How and why was an organization for the *izoku* established? What are Japanese attitudes to death and religion? How are the JL123 victims remembered? What do the *izoku* do now for the anniversary?

Part IV turns to the narratives about JL123. Chapter 7 analyses the media's coverage of the crash, both in 1985 and the annual anniversary events. How was the crash reported? How has the passenger and crew list been dealt with? How did the media get its stories? How does the media report on and visualize death? To what degree can one see differences between the main newspapers' reporting of JL123 in August 1985? What difference is there between national and local news reporting? How does the media cover the anniversaries of the crash? Chapter 8 looks at why, 25 years since the accident, there is still so much interest about what happened both in Japan and globally, and also considers the books and films relating to the crash. What do we mean by history and social memory? Why are there conspiracy theories surrounding JL123 and what are they? Why did so many people report other coincidences and 'unexplainable' experiences? What sorts of books have been written relating to JL123? Why have some novels been written? Why have these novels been dramatized and in what ways is the content different to the novels? Why is there global interest in JL123? What are the differences between the reasons for Japanese and non-Japanese peoples' interests in JL123? What are the causes and implications of these differences? Why do non-*izoku* also visit the crash site?

In Part V, Chapter 9 draws together all the key points that have been raised throughout the book and considers how JL123 may be remembered in the future. What was the cause of the crash? What lessons can be learnt from the way the remains were identified and the *izoku* were treated? Are we likely to see an increase in interest in JL123 in the future? How should JL123 be remembered? Ultimately, the book concludes that JL123 is Japan's and the aviation world's equivalent of the sinking of the *Titanic*. It is a story which is likely to garner much interest, but one where certain mysteries will always remain, which itself may fuel some of the continued interest.

In answering all of the above questions, the focus will be on what the answers reveal about Japanese society and, where appropriate, other contemporary societies around the world.

Why study JL123?

This book deals with why there is interest around the world in JL123, but I believe I should set out why I chose to write this book and to write it in the way I have. I originally had no plans to write this book and had just carried out some initial research for a different project when a series of coincidences came together. Eventually it reached a point that I felt a calling to write this book.

The seeds to create this book were sown in 2005. I had arranged to meet with Robert Rigby and Keith Haines from JAL's London office. Keith was running late and Robert had to leave at about the time Keith arrived. Just before Keith arrived, Robert informed me that Keith had been involved with JL123 and that I should ask him about it. I did so. It turned out that Keith had been the JAL employee responsible for looking after the family of the sole British victim (excluding those on a British–Hong Kong passport), Kimble Mathews. Keith offered to come and talk about his experiences at a seminar at the University after his retirement in September 2006. The date was subsequently set for 20 March 2007. In December 2006 Keith suggested that Peter Mathews, Kimble's father, also participate. Keith approached Peter, who agreed to the request. Whilst Peter and Keith could speak about their personal experiences in the days following the crash, I felt that an introduction covering the details of the crash itself would be needed as I was aware that some, if not most, of the audience would have little, if any, knowledge of this and that without knowledge of the crash itself the experiences of Peter and Keith would lose their impact. It was agreed that I would do this introduction.

When I began looking into JL123, I became aware of the lack of writing in English on the subject. There are no books in English solely on JL123. Indeed, even among books on aircraft disasters, chapters on JL123 are relatively rare, but I came across a number of websites and, more surprisingly, videos on YouTube – both of TV footage from 1985 but also of where people had recreated parts of the flight on computer software. Why was this happening? I could see there may be a possibility to write a book using JL123 to study different aspects of Japanese society and culture, just as I had done with my previous book on the shinkansen, whilst also looking at the global interest in the crash.

But there were other reasons why I began to feel that I should write this book. JL123 was the first time I remember seeing Japan in the news. I was actually on holiday in France on the day of the crash, but I saw the headlines about the crash in newspapers that evening when going out for dinner. Upon returning to Britain a few days later, JL123 was still one of the major news stories. I have vague memories of seeing some of the reports, although the only one that left an impression was the theory that the crash was due to a broken door. I also remember discussing the crash with friends when the school term started again in September. I remember this distinctly as the mother of one friend told me how she had a premonition about the crash.

My next memory of JL123 was at upper secondary school in about 1988 when it came up in conversation with some Japanese friends. I remember commenting on the broken door, and although they suggested that this was not the cause they could not remember what it was. Between 1993 and 1994 my wife and I spent a year on the Japan Exchange and Teaching (JET) Programme. My wife was located by the Japan Sea coast in Toyooka, although the *onsen* resort of Kinosaki (subsequently subsumed as part of Toyooka as part of the programme of city mergers) was her nearest population centre. She became acquainted with the owners of a *ryokan*, whom I also met on a number of occasions. Although I do not remember how the conversation arose, it transpired that the owner's wife's father was on board JL123, and we briefly discussed the cause and her own actions after the crash, including writing a song whilst she was at college. Although I touched briefly on the JL123 crash in a lecture in one of my courses at university, my next memory of discussing JL123 was not until April 2003 when I learnt that a friend could have been on the flight, but due to his love of trains, had instead decided to travel to Ōsaka by sleeper train with some work contacts. He did not learn about JL123 until after his train arrived in Ōsaka the next morning.[1] Later that same year, on a company visit doing fieldwork for my book on the shinkansen, I travelled with an employee from JR Tōkai, who was reading a book about JL123, and we discussed the crash as we travelled. Then, when I first met Peter Mathews in 2007 and mentioned that I was considering writing a book, he said that I could use the diary he wrote and photographs he had taken in August 1985. I realized that I had access to resources which many people could not possibly have and felt that I had to write this book.

Perhaps there were other factors also guiding me to write this book. Out of the books I read as a child, there are only two I particularly remember: *Sam 7* by Richard Cox, a fictional story about a plane crashing on London Victoria Station, and *A Book of Disasters* by Jane Ferguson. The second book, published in 1981, covered many types of disaster, but the section on earthquakes and entry for *HMS Hood* were of most interest to me. I cannot say that reading about the Great Kantō Earthquake at age ten did much to spur my desire to get into Japanese studies, but it may help to explain why I have subsequently considered writing about earthquakes in Japan.[b] I also enjoyed watching documentaries about plane accidents – particularly around the time I was doing my PhD, as I respected the way that the research into finding the cause was done – although this, together with other experiences,[2] may have been one reason why I gained a fear of flying around this time.

Some have asked me why do research which might involve criticizing Japan? This was an issue I never considered when I started the research, although I was aware of the premise for the question and was aware that there would be criticism. Ultimately I think the answer to the question stems from a desire to understand Japan and Japanese society in its entirety. People who study Japan do not necessarily like everything about the country, but most of us would like to see it learn from its mistakes and perhaps we have a role to play in helping Japan learn about itself as much as we are trying to learn about it and teach other non-Japanese people about Japan.

Other than the personal reasons for studying JL123, there are, of course, larger issues. From an academic point of view there is much to be interested in. The crash and subsequent events can be used as a tool to study Japanese society, and to some extent global society. That it is appropriate to do so can be seen in the context of studying Japanese 'culture'. Culture is a term, as discussed in my previous study,[3] that continues to be a source of much debate. I do not wish to go over that debate again here, but merely to note that many definitions of culture, such as Kluckhohn's,[4] often refer to how society responds to problems.[5] A disaster clearly constitutes a 'problem', so in trying to understand Japanese culture*s* – for speaking of a single Japanese culture is too simplistic – the study of JL123 would appear to be an obvious means to do so.

Kyoko Hatano 波多野京子 (24) · Mikiko Ōno 大野美紀子 (26) · Seiko Ōno 大野聖子 (24)

Conducting and presenting the research

I believe that it is appropriate for me to make clear some of the assumptions made in writing this book, including some information about my own background and views on subjects covered in the book. The reasoning, without going over again at length what I have discussed previously,[6] is that as it has become the norm in academic writing to write in the first person, I believe it is necessary for the reader to have some idea of who that 'I' is. My views on certain issues will be introduced as appropriate throughout the book.

In the broadest sense, I have assumed two main readerships for this book – those interested in Japan and those with an interest in aircraft disasters. These two groups need not be mutually exclusive, but, given the lack of publications such as this, it would seem the two groups have little overlap, and the level of knowledge in each group may be varied. Even among those interested in Japan there may be many who do not have detailed knowledge of Japan; for example mentioning a politician's name may not conjure up any particular image or be supported by any significant knowledge of that politician's agenda, let alone a significant understanding of the Japanese political system. To ensure readers can appreciate the text, a degree of contextual explanation is necessary. The risk, of course, that too much of this will reduce the amount of new information and analysis which can be provided and reduce the overall impact of the book. It is a difficult balancing act, particularly when dealing with a subject where there are two such, potentially diverse, readerships.

Of course, it should be stressed that even those employed in Japanese studies, which usually refers to those researching some aspect of Japan and Japanese society rather than the Japanese language, are not experts in all aspects of Japan. Japanese studies covers a wide range of subjects – including political science, sociology, business, anthropology, linguistics, literature, urban studies, religion, history, media studies and so on. Many of these subjects are studied in almost complete isolation from each other, and, all too often in isolation from those in similar fields who do not study Japan. A book such as this, which has chapters dealing with a number of the above list, has the potential to be of interest to a wide range of Japanese studies experts. Some will merely read the section relevant to their speciality in order to find new information and ideas that can be used to help in the development of their own research, whilst others will read more, and be touching upon areas with which they are less familiar and so may need nearly as much 'background' information as the reader unfamiliar to Japan.

Within Japanese studies, there is one significant group that cannot be described as being 'experts' in all aspects of Japan yet. These are the students we teach. It is too much to expect, depending on the structure of their degree and the number of subjects/courses/modules they study, for them to have an in-depth knowledge of all areas of Japanese studies, let alone the more universal theories used within the fields which are covered. However, I believe that these students need to know a little bit of everything relating to Japan. When they meet a stranger on a train, for example, it is not uncommon for a range of questions to be asked relating to a variety of contemporary and historical issues about Japan, the language and its people. Someone who has merely visited – or lived – in Japan may not be expected to know the answer to such questions, but someone who has studied Japan should be able to answer, even if the question relates to something beyond their usual scope of interest. In that respect a book such as this, which covers a wide spectrum of subjects under the banner of Japanese studies, can be a useful addition to the students' reading list.

Returning to the two readership groups, there is another group which overlaps those – the Japanese readers. Many Japanese take great interest in what non-Japanese think of them and their country, and on top of this, there is an awareness that non-Japanese can often access information and people which the Japanese themselves may have trouble accessing. The Japanese seem particularly sensitive to this, perhaps due to the experiences of *gaiatsu* (external pressure) in bringing about change. Of course, for the Japanese readers many of the fundamentals of Japanese society, which it is necessary to point out for the benefit of some readers, are norms to which they well accustomed and so the information may appear of little relevance. However, I hope these Japanese readers will be aware that the need to point out this 'norm' is an indication that this 'norm' may be indicative of something different about Japan (in relation to my own culture, at least), and encourage them to consider, where it is not done in the text, the reasons for this difference and what it represents and symbolizes.

What is perhaps a little unusual about my previous book and this one, is the expectation that the book may reach beyond the academic community. Taking that into consideration, combined with the thought that some may be unfamiliar with the nature of British academic writing and/or Japanese studies, I would like to explain how this research was done and how the assumptions and conventions (summarized in Notes on style) were reached.

One of the critical first steps in embarking upon new research is to secure funds for the research. In the case of this book, it is highly possible that I would not have been able to conduct this research had it not been for the existence of 'Individual Research Budgets' within Cardiff Business School, of which the Cardiff Japanese Studies Centre was a part when I was conducting the research for this book. This budget can be used on anything appropriate to research. I used some of these funds to order a shipment of books about JL123 from Japan. It was through acquiring these books in January 2007 that I was able to conceptualize what the focus of my research could be. In an area such as Japanese studies where large sums for research – particularly for the acquisition of books – can rarely be obtained, a means which enables a new piece of research to begin as well as to develop is invaluable. Most funds in Japanese studies tend to be for fieldtrips. The limited sums available to the various foundations has an impact on what can be funded and the sums which are available, so that in the end it becomes natural – whether a formal limitation or not – that applications are made solely for fieldwork (i.e. the cost of transport to and within Japan and subsistence within Japan) and do not include applications for materials (including books, recording devices, etc. which researchers often need). When applying for funds from organizations, one standard question is to evaluate the importance and impact of the research. Even for a subject such as the one covered in this book it is hard to suggest that there will be significant impact. Luckily, as detailed in the Preface and acknowledgements section of this book, I did get support for this research from organizations with a particular interest in Japanese studies. One can only hope that as our societies continue to develop that greater respect and appreciation will be given to the academic community to allow greater sums to flow to all fields of research (and teaching) so that we can fulfil our potential.

Fieldtrips formed an integral part of this research. The main elements of the fieldtrips included visits to companies or organizations, interviews and observations. Naturally there was a degree of overlap between each of these. Such fieldtrips have their limitations. For example, in relation to interviews, although a popular form of research, one has to keep in mind the perils inherent in them. Interviewees may not be representative, for example. In a country such as Japan, where relationships tend to be made through established connections rather than by 'cold calling', studies can lead to interviewing similar-minded people which can distort the results of the research.[7] Interviewees may also not be able to express what they really think, or they may even prefer to express what they would like to be recorded as the 'truth' compared to the real reason. Interviewees may also try to rationalize the irrational.[8] In the case of this research there is an additional problem in that questions were sometimes dealing with events about 25 years previously and so time, let alone the maelstrom of events at the time, could further confuse answers. In the end, I tend to prefer to use interviews as a background and to add colour to the information gathered from other sources, keeping in mind the limitations inherent in those sources, rather than using them as the cornerstone of my research. Interviews can be enlightening, but are rarely remarkable. At the extreme, as I found with some of my previous research, an interview with a previous Prime Minister is largely predictable because only limited questions, i.e. ones they are prepared to answer, can be asked, and the answers to these are likely to be already known – particularly if they have written much, or much has been previously written of them. Sometimes, however, interviews will present information which will be the spark for a new piece of investigation, although the interview itself does not subsequently get referred to in the results. Some interviews can be extremely beneficial and so the time-consuming process is not without merit. Most of my interviews were semi-structured, whereby some questions were pre-prepared and asked in order, and additional questions were often added as appropriate. In some cases questions were sent in advance so that time could be given to prepare answers.

The most significant problem with observations is whether one can correctly interpret what is being observed. Observations need to be supported by further interviews and reading about the subject. In this respect I have found that multiple observations of the same event can be beneficial as it provides an intervening time during which further investigation can be done, and then these ideas can be tested upon a future visit. Follow-up interviews can be beneficial for the same reason. Additional observations, at another time, also allow for comparisons to be made.

Consequently, I find relatively short, but numerous, fieldtrips tend to be effective.[c] A further problem with observations is that they are most effective when one can act as a 'fly on the wall'. However, it can be hard achieving this, particularly as a foreigner in Japan (see Chapter 7). Another issue with observations in relation to this research was the time taken to do them. For example, the study of the memorials at Osutaka-no-One took several hours – with several hundred memorials, spending only a minute at each memorial would mean taking a whole day, and often more than one minute each was needed. This time-consuming piece of research was necessary, but resulted in only a few pages in Chapter 6. A further problem with observations and interviews is getting the balance between describing 'the norm' without ignoring exceptions, but also trying to ensure that descriptions of exceptions do not over-emphasize their significance. Stories of the war – particularly those portrayed in films – tend to focus on the action, whereas some of those who were involved in the conflicts prefer to emphasize the relative dullness and routine nature of what happened day to day.

In conducting interviews and observations there are also ethical issues to be considered. Throughout my research I worked on the basis that I would not interview any *izoku* unless they approached me first and agreed to be interviewed. I felt that they had suffered enough and it would not be appropriate for me to make them go back over painful experiences. My research plan was based around getting information about the *izoku* in other ways and so their direct comments would not be necessary. Similarly, I did not try to track down the survivors and interview them. As with the *izoku* (and indeed, three of the survivors are *izoku* themselves), I felt it would be inappropriate to make them recount yet again what they had experienced. But my interviews and observations, as with Reader's research about the pilgrim community, was not about the individuals themselves but as representatives of the community.[9] Although I have listed the names of many people who I interviewed in the Acknowledgements, I interviewed and observed many more. There were many whose names I did not need, or sometimes I decided it would be better to maintain their anonymity and so their name is not listed. Similarly, unless relevant, citations within the text are often anonymized.

Another piece of research for this book was an online survey. Surveys are often even more popular with students than interviews, to the extent that some appear to think that original research equates to conducting a survey. I do not trust surveys, but think that they can, like interviews, be a useful tool to add colour to rest of the text. As with a previous survey I did,[10] I have come across cases where it was clear that the respondent was providing false information. For my JL123 survey it was hard to design questions which would highlight a probable 'false' submission as I had done in my previous survey. That does not mean it was impossible to uncover them altogether, but it was more time consuming. For example, I had one case where the person agreed to be contacted by email so that I could ask more detailed questions. It was only in doing so (and by looking for further information about this individual on the internet) that I discovered that my suspicions about their lie were confirmed. Whereas it was then possible to exclude that individual's submission from the results, it is impossible to know how many other inaccurate submissions there have been. One can only hope that as the sample size grows, the number of deliberately misleading submissions will drop to a level where it has little impact upon the results. Why people would follow a link to take part in a survey to deliberately mislead results is puzzling to me. Perhaps it was a desire to seem important through their answers. Whatever the case, my impression is that sometimes an individual's moral standards on the internet may be lower than that in 'ordinary' life and that participation in online surveys and false responses is just one manifestation of this. A further problem I encountered was people trying to take advantage of my access to information and people, often by providing me with false information in the process.

This research called for the reading of a number of books. However, particularly when starting research on a new topic, it is likely that the first book read is likely to be highly influential as so much of the information will be new. Each subsequent book is likely to contain less new material and so the amount of text marked up as being significant will be less. Consequently, the order in which books are read may have an undue influence in the number of citations from each book, or the way in which conclusions are reached. As with interviews, we need to keep these issues in mind when thinking about the direction research takes. When thinking about how much information to cite from a book, you soon appreciate how much time can be wasted on spending too much time on any one book. If reading slowly and in detail, marking the new (significant) text, a book can take weeks to complete. But if I were to take just 100 words of text from each chapter of each book I read for this research, that would amount to more than the 100,000 word length of this book. There would be no space for my own findings or observations. There is no need to try to remember every word, every sentence or every page. This is the basis of speed reading.

One of the factors that has aided me in the reading for this book is the significant use of Japanese materials. For large parts of the research, there was no English material available, so using Japanese materials was not only desirable but unavoidable. Presenting some of this information is one of the significant roles of those of us in Japanese studies. So long as our findings are used more widely – in other words, by those outside Japanese studies also – we will have helped to add to the body of literature. That we are not merely translating but also presenting the results of our own research is another of our roles. Although I have previously argued that knowledge of Japanese is necessary to understand the nuances that can be 'lost in translation',[11] I am no longer convinced that this is the case. I doubt that few non-Japanese speakers of Japanese actually have the level of language ability to understand all of the subtle nuances, etc. that can exist; few ever reach a stage where their Japanese is 'perfect' (though one has to remember that most are also not 'perfect' in their own language). In many cases, I would also argue that a high level, let alone 'perfect' understanding, of the foreign language is not needed. This is particularly the case when we take into account how many Japanese people speak English. It would be arrogant to suggest that a Japanese person cannot explain in English something as well as a Japanese studies specialist with Japanese language ability can comprehend a Japanese explanation.

I would argue that the dangers of not interviewing enough people or reading enough are far greater than any problems posed by being reliant on an interpreter/translator (and a good translation should include contextual information as well as the translation of the text itself). For sure, not having any ability in Japanese should not be a barrier to studying the country – that Bathes' *Empire of the Signs* is seen as such a classic is perhaps a demonstration of this point. The obstacles that such people face should not be insurmountable and I certainly do not believe that Japanese studies should be building barriers between those who can and cannot understand Japanese. If those of us in Japanese studies who can understand Japanese have presented our findings in a clear manner, then this should be a means to help those without Japanese ability (and translators) to use their skills to research Japan. Without this, Japanese studies is likely to remain overly divorced from other academic disciplines and unlikely to reach its full potential.

In my previous book, I discussed the problem of conducting research and the possibility of 'burning bridges' or presenting material which would be 'problematic' for the private companies which assisted me with that research.[12] In this research it is impossible to separate the crash from JAL. It would have been possible for me to conduct this research without JAL's assistance, but having their assistance has undoubtedly been highly beneficial. As will be discussed elsewhere, it is perhaps fortunate that I have been conducting this research at a time when JAL has been dealing with the crash in a more open manner. During the course of this research I have got to know many of the people which this study is about. On top of this, as discussed in other chapters, I have become personally attached to this disaster and its long-term impact. Indeed, there is sometimes a question mark whether the purpose of research is to gain a greater understanding of the target of our research, or whether it is to gain a greater understanding of ourselves. At times it has been hard to separate what was required for this research and what I wished to do on a more personal level. As someone who flies quite frequently, it is perhaps natural that I should be concerned with the promotion of air travel safety. As a combination of these reasons I set up a group relating to JL123 and air safety on Facebook.

Conventions

Let me now explain some of the conventions used in this book – summarized in the Notes on style. As the readership of this book may not only cross academic disciplines (both inside and outside Japanese studies) but between academia and a more general audience, I believe it is necessary to highlight some of the issues. In deciding on some of these conventions I have had to question some of the ways in which I have been taught, and been teaching, since I started studying Japan in 1989. As with my comments about Japanese language ability in studying Japan, I am conscious that my views may not be appreciated by all, but I believe that a discussion is necessary as I think that some of our conventions are no longer, if they ever were, appropriate and are maintained due to custom rather than for any logical reason.

Let us start with issues concerning the Japanese language. Japanese is a phonetic language, with five basic vowel sounds. Vowels may stand alone or be used in combination with certain consonants or vowels. All Japanese words are made up of these sounds, and upon hearing the sound it is possible to write down the appropriate character from the Japanese phonetic alphabet, *hiragana*. It is possible to dictate the Japanese Constitution to students after their first week at University and they can write it down in *hiragana*. Of course, normally it would be written in combination with *kanji* (Chinese characters), but the point is that sounds are unchanging and are not linked to meaning. Compare that to English, where students would not know whether I meant 'right' or 'write' if the word was said out of context. Despite the strict nature of Japanese phonetics, that does not mean that there are not problems, as will be discussed elsewhere.

Japanese words can be Romanized. The standard system, a revised version of which is used in this book, is referred to as the Hepburn System. The system includes macrons on long vowels. However, it has become a convention in Japanese studies that certain words do not have macrons. So, rather than writing Tōkyō (or Toukyou) as it should be, it is normal to write Tokyo. But where do we draw the line? When do we decide, for example, that Kōbe can be referred to merely as Kobe? What have we achieved by this alteration? Surely is it anti-educational as well as being technically incorrect. Either we accept that readers do not understand macrons and will not use them themselves and so we do not use them at all, or we maintain all of them to preserve the correct form. The halfway house makes no sense and should be stopped. I also believe that English should use local spellings for place names. It should be Kyiv not Kiev, Napoli not Naples, and Firenze not Florence. Why is it that Lecce is Lecce, but Livorno is Leghorn? It would be nice to think that local pronunciations could be adopted too – Tōkyō rather than Tokio and Pari rather than Paris, but while we have complications within Britain with our own towns (e.g. the pronunciation of Shrewsbury), this may be too much to expect.

There are problems with the Romanization of Japanese in that, without the knowledge of how Japanese vowels are pronounced, Japanese words, even if Romanized properly, may not be pronounced properly. Similarly, words that look like English words can cause confusion, although the additional use of accents (not used in standard Japanese Romanization) can help – so the drink saké may be preferable to merely 'sake' (which need not be italicized as it appears in the Oxford English dictionary), which could be confused with the English word 'sake'. In this book, the problem word is '*one*', meaning ridge. Although the italics can help to distinguish it from the English 'one', I suspect that few non-Japanese speakers would think to pronounce it as if it were written 'oné'. To resolve this possible problem, and to maintain the standard Romanization, I have decided to not use the word in isolation, as in the text it is merely used when referring to the location of the JL123 crash – Osutaka-no-One (ridge of Osutaka).

Another problem with dealing with Japan is name order. In Japanese the family name comes first, then the given name. In Japanese studies, the norm is to maintain this – although Americans of Japanese descent, for example, use given name then family name. Personally, I do not have time to check an author's genealogy to see which is correct and it is amazing how many times it is not obvious as far too many publications still do not make their conventions clear. I have always used this convention, but in conducting this research I came to question the legitimacy of doing so. Can we expect people to know conventions for each culture or country? Do academic texts have a responsibility to educate and so maintain local customs? A book or journal should have a 'Notes on style' section which makes its conventions clear – although there is always a risk that text will be read without reference to the 'Notes on style'. But will all readers able to spot which name is Japanese, Korean, Chinese or Indian? Whilst I have few problems with the first three, I was not aware until doing this research of the multitude of systems used in India. This is pertinent as there were Indian victims on JL123. How could I be sure that I got the ordering of these individuals names correct? The potential is, in maintaining the 'correct order' (i.e. how it would be ordered if written in their native language), that the text becomes unnecessarily confusing and cumbersome. Unless we are to write family names in capitals, for example, which is not pretty, my only conclusion is that we should use the convention of the language being written in, and it is for that reason why I have decided to abandon the conventional custom in Japanese studies.

Let me now discuss references. References are provided to show where information came from. The analogy is a lawyer presenting evidence in a court case – merely saying there was a crucial letter would not be sufficient without showing the letter in question. Whether a citation/quotation (or the whole letter) was then also used is a separate issue. References and quotations are also different from notes, which are a means of providing additional information which is not central to the main text, so need only be read if the reader has a particular interest in this part of the text. These may seem obvious points, but it never ceases to amaze me how even post-graduate students confuse them. It has seemingly become a particular problem in the internet age when information is often presented without referral to where it came from. It should be noted however, that there is a cultural bias here. Many Japanese books include little or no referencing, no bibliography and no index – although I would say that many of these books contain a higher percentage of one's own work than may be found in British academic works. Referencing can follow a multitude of systems. A popular one is the Harvard system – and many others are variations on it. Within this system, which I used in my previous two books, the text and page being referred to are placed in brackets at an appropriate point in the sentence. For example '(Hood 2006:1)' would mean page 1 of the publication, listed in the bibliography by Hood, published in 2006. A letter after the date is used when there are multiple publications in one year by that author. The major problem with this system is that the reference gets in the way of the main text and is distracting. Whilst this may not be such a problem for academics, who are accustomed to it, for the general reader it can be particular annoyance. I received a significant amount of feedback from railway enthusiasts who read my previous book to this effect. The problem is that there is no printable version of the small dialogue boxes which appear when you hover the mouse over a reference in a word processing program.

An alternative system places the reference text with the notes, with the use of superscript numbers in the text to denote when a reference or note is being made. The problem with this system, as generally notes and references are not put at the foot of pages but at the end of the chapter or in a separate section after the main text, is that when encountering one of these numbers the reader cannot know whether turning to the appropriate part of the book will provide them with a reference or a note. Whilst academics may love references, there is not a need to check each one. It is a means to provide a link to further reading on a subject which the reader wishes to know more for some reason. It can be frustrating to turn to the appropriate page expecting to find a note, only to discover a reference. The referencing system in notes also tends to use 'Op. cit' (Latin, short for 'opus citum' and used when referring to texts referred to elsewhere in the references) and 'Ibid' (Latin, short for 'ibidem' and used when referring to text mentioned in the proceeding reference) to abbreviate the references.

So there are problems with both of these systems. For this book I took elements from both of them. To overcome the unsightliness of the Harvard system I have put references in note form using a superscript number. However, the reference itself uses the 'Harvard style' rather than the 'Latin style', which I find tedious. To save space (which is a concern for the cost-conscious publisher), these references can then be laid out in columns in a 'Reference' section after the main text. To distinguish notes from references, superscript letters are used to denote notes (which in most texts should be relatively minimal in number) and are placed in a 'Notes' section after the main text.

Having set out how the research was conducted and the conventions which were used to present the research, we can now move on to put JL123 in context.

2 Putting JL123 into context

> The year 1985 was overwhelmingly one of Kansai. Beginning with the release of the composite sketch of the Fox-Eyed Man, it was a tumultuous twelve months that saw the outbreak of the Yama-Ichi War; the crash of a Japan Airlines jetliner carrying a lot of people from Kansai, including President Nakano of the Hanshin Tigers baseball club; the live telecast of the fatal stabbing of Chairman Nagano of Toyoda Trading; and, most unbelievable of all, the Hanshin Tigers in the race for Japan's professional baseball crown.[1]

For many non-Japanese I suspect most of the events described above are meaningless. For many Japanese of a certain age they may find little resonance. Yet, these words by the infamous 'freelance *yakuza*'[2] Manabu Miyazaki, who was thought to have been the 'Fox-Eyed Man', are probably a fair summary of 1985 Japan. It is all a matter of context. The need and importance of getting the context right struck me when I first visited Ueno-mura. Having read a number of books on JL123 I was surprised that, when I first visited, the reality of the layout of the crash site and its location did not fit with what I had expected. These geographical issues will be dealt with in subsequent chapters.

Flight JL123 crashed in Japan in 1985. There may be readers who are not familiar with Japan. Some readers may have no memory or experience of life in 1985. Some may not be familiar with the nature of the Japanese transportation market. Some may not be familiar with the history of JAL. Some may have little knowledge of disasters, particularly within a Japanese context. I suspect many readers will lack a combination of all of these aspects. It is not possible to understand the JL123 crash and the responses to it without some understanding of these areas. Much of the 'Japan' context will be introduced as appropriate throughout the book, but I believe that understanding the other areas are of such significance that it warranted its own chapter rather than being part of the first chapter.

1985

1985 may not have the same ring to it as 1984, with its social and political overtones thanks to the work of George Orwell, which has given the world terms that have become familiar concepts and names of television programmes. However, for some, 1985 was a very significant year. In Japan the year perhaps had extra significance, as rather than being known as '1985', it was the sixtieth year in the reign of the Emperor – 'Shōwa 60'. Round numbers tend to be particularly significant anniversaries, and with 1985 the fortieth anniversary of the atomic bombings of Hiroshima and Nagasaki, as well as the end of the war itself, there was extra poignancy.

For many of my current students, 1985 is as much a part of history as the early 1960s were for me. A world in which mobile communications were largely non-existent may be as alien and unimaginable to them as a world without colour televisions in most homes was to me. It is far too easy to misunderstand the realities of the past when viewed from the world of modern comforts and conveniences.

The politics of the 1980s were memorable due to there being strong, charismatic leaders in many of the leading nations of the world – the G5 as it was then. The president of the United States was Ronald Reagan, having just won a landslide election victory for his second term in office. The British prime minister was the 'Iron Lady' Margaret Thatcher, who was enjoying high levels of support after her party's landslide victory in 1983. Thatcher built a strong bond with Reagan, cementing the so-called 'special relationship'.[3] Also developing a special relationship with Reagan was Japanese prime minister Yasuhiro Nakasone. Although he had not enjoyed widespread support when he first became prime minister, by 1985 he was popular, partly due to the 'Ron-Yasu' relationship which was symbolic of a more internationally engaged leader and Japan.[4] In West Germany, as it was then, the chancellor was Helmut Kohl, who in 1985, was in the fourth year of his incredible sixteen-year tenure. Not only did he oversee German reunification when the Cold War ended, but he helped mastermind the Maastricht Treaty, which set the foundations for the modern European Union and the Exchange Rate Mechanism which would lead to the creation of the Euro.[5] Another person central to this process was François Mitterrand, who became French President in 1981 and remained in position for a record fourteen years. In March 1985 Mikhail Gorbachev became the general secretary of the Communist Party of the Soviet Union and would, in time, herald a new era though the introduction of *glasnost* ('openness') and *perestroika* ('restructuring'). These leaders helped to bring the Cold War to an end, but until that came about I remember being apprehensive about their brinkmanship. The arms race was still raging and the number of nuclear missiles the United States and Soviet Union possessed was said to be enough to blow up the world several times over. I fully expected to see a nuclear mushroom cloud during my lifetime. Programmes set around what could happen during such a war, such as *Threads* (BBC, UK, 1984) and *The Day After* (ABC, USA, 1983), as well as anti-war songs such as *Two Tribes* (Frankie Goes to Hollywood, 1984), *99 Luftballons* (*99 Red Balloons*, Nena, 1984), *Pipes of Peace* (Paul McCartney, 1983) and *The War Song* (Culture Club, 1984) struck a chord.

In terms of music, 1985 was the year of Live Aid, with many of the world's top international artists coming together in London and Philadelphia, with six smaller concerts held in other cities globally, to help raise money for those impacted by a severe famine in Ethiopia. Whilst the names of many of those who performed would be recognizable to the modern youth, there are many others which hark back to a different era in pop music. Hit songs in 1985 included *Into The Groove* (Madonna), *19* (Paul Hardcastle), *The Power of Love* (Jennifer Rush), *Take On Me* (A-ha) and *I Want To Know What Love Is* (Foreigner). It was a time when CDs had only recently begun becoming popular; vinyl was still the main means of buying albums and singles. Many people recorded tracks onto cassette tapes so that they could be listened on personal stereos – with Sony's original Walkman having only been introduced a few years earlier and the CD-version ('Discman') in 1984. Popular songs in Japan in 1985 included *Friends* (Rebecca), *Roppongi Chūshin* (An Ruisu), *Julia Ni Shōshin* (The Checkers), *Mi Amōre* (Akina Nakamori) and *Romantic Ga Tomaranai* (CCB). On the big screen, the hits of the year included *Back to the Future* (Robert Zemekis), *The Color Purple* (Steven Spielberg), *Rambo: First Blood Part II* (Sylvester Stallone), *Rocky IV* (Sylvester Stallone) and *Out of Africa* (Sydney Pollack). In Japan, some of the top films were *Biruma No Tategoto* (Kon Ichikawa), *Tampopo* (Jūzo Itami), Be-Bop High School (Hiroyuki Nasu), *Ran* (Akira Kurosawa) and *Otoko-wa Tsurai-yo Tora Jirō Renaijuku* (Yōji Yamada), the thirty-fifth of the 48 main series of the popular series. Although computer games were still in their infancy, 1985 saw the release of the original *Super Mario Brothers* video game.

On the sporting front, Ayrton Senna won his first Grand Prix, whilst Alain Prost won the F1 Championship. In football, Everton were crowned First Division Champions ending Liverpool's three-year reign. In the European Cup Juventus beat Liverpool at Heysel Stadium, Brussels in a match best remembered for the loss of 39 lives during violence prior to kick-off. In tennis, the year started with Martina Navratilova winning her one-hundredth tennis tournament before winning at Wimbledon for the sixth of her nine single championships there, whilst a 17-year-old Boris Becker became the youngest and first unseeded player to win the men's championship at Wimbledon. For snooker it was the year when more than 18.5 million watched until after midnight to see Dennis Taylor beat Steve Davis in a remarkable final of the World Snooker Championships at the Crucible in Sheffield.[6] In American Football, the San Francisco 49ers beat the Dallas Cowboys in Superbowl XIX, whilst the Kansas City Royals beat the St Louis Cardinals in the World Series. In Japan, the Hanshin Tigers beat the Seibu Lions to win the Japan Series baseball championship (see Chapter 6), and PL Gakuen won the Summer High School Baseball Championship. In Sumo, the great Chiyonofuji won four of the six tournaments, two of them without any defeats during the 15-day tournaments.

By 1985 Japan's economy was going from strength to strength. Although its initial economic expansion in the 1950s and 1960s had been domestically driven, exports became increasingly significant. By 1985 exports accounted for 13 per cent of the economy and the trade surplus was ¥11 trillion.[7] In response to this, only a few weeks after JL123 crashed, the G5 leaders met and signed the Plaza Accord.[8] The aim was to depreciate the US Dollar in relation to the Yen and the German Deutsche Mark, to make American exports cheaper which should have boosted sales of American goods overseas and reduced the purchase of Japanese and German goods in the US. However, as the Japanese economy continued to strengthen, together with a belief in 'the Japanese management system' being superior, so the resulting confidence may have helped to paint over other problems in the economy and business operations. The property speculation that went with this led to the creation of the 'Bubble Economy', which eventually burst after the realization that 'facts' such as the land around Imperial Palace being worth more than the whole of California[9] was nonsensical, so leading to a period known, inappropriately in my opinion, as the 'Lost Decade'.

Although Japan was a heavily developed and modernized country by 1985, only 42.8 per cent of homes had a microwave (96.5 per cent in 2004), 52.3 per cent had an air-conditioner (87.1 per cent in 2004), 27.8 per cent had a video recorder (82.6 per cent in 2004) and 67.4 per cent had a car (86.0 per cent in 2004).[10] Furthermore, 19 per cent of homes were still without a telephone,[11] with ownership lower in rural areas. Some homes had a telephone in their *genkan*[a] and a box where people could leave money as a contribution to the phone bill. There were over 900,000 telephone boxes around the country.[12] Phones were coloured according to what type of call could be made from them and what form of payment could be accepted. Mobile phones were only just being developed, and although the first mobile phone call in the UK was on 1 January 1985,[13] Japan had started the world's first mobile service in 1979. But in 1985 most devices in Japan were large and connected to cars, for example.[14]

Referring back to the Miyazaki quotation at the start of this chapter, 1985 saw the conclusion of the Glico–Morinaga bribery case which had been gripping the nation for 18 months. The first incident, the kidnapping of the Glico president had taken place in March 1984. There were subsequent extortion attempts against the company, and then against Morinaga, Marudai Ham and House Foods. The prime suspect called themself the 'Fox-Eyed Man' or 'Monster with 21 Faces'. On 12 August 1985 a final blackmail message was sent saying that the campaign against the companies was ending. The case remains unsolved with the statue of limitations having passed.[15]

Many Japanese remembered 1985 as the year of the Tsukuba Science Expo, when Japan showed off various technologies and their uses, as well as hopes for possible future developments and uses.[16] Although not on the scale of the 1970 Expo in Ōsaka, which attracted 64 million visitors,[17] the 1985 Expo still had an impressive 20 million visitors.[18]

Two symbolic discoveries were also made beneath the waves in 1985. On 3 August it was announced that the wreck of the *Yamato*, pride of the Japanese Imperial Navy, had been found.[19] Then, on 1 September, Robert Ballard's team became the first to discover the wreck of the *Titanic*, 73 years after its sinking.

Transport in Japan

Let us now put the Japanese transportation market into context. Figure 2.1 shows the modal split depending on the length of journal for fiscal years 1983 and 2005. Road and rail are favoured for short and medium distance journeys, and as the

Yōko Fujikura 藤倉洋子 *(16)* · *Tomoko Fujikura* 藤倉智子 *(13)* · *Masahiro Fujimoto* 藤本正裕 *(23)*

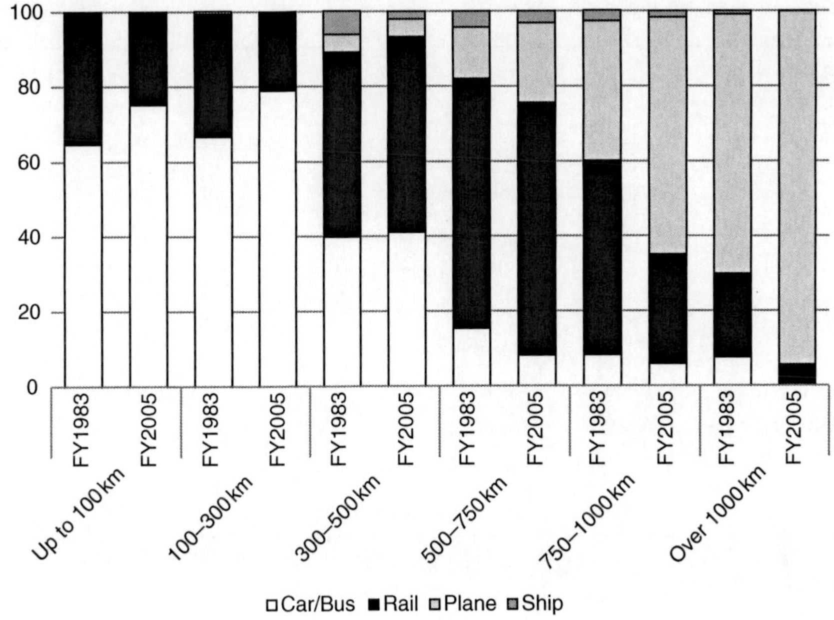

Figure 2.1 Modal share of transport in fiscal years 1983 and 2005

Figure by the author. *Sources:* MLIT 2010c, MLIT 2010d.

distance increases so the usage of planes increases. One can also see that there has been a shift in usage, which not only reflects the availability and improvements in different modes of transport, but also responses to problems associated with those modes of transport.

The population of Japan is about 127 million, about 40 per cent of that of the USA and 25 per cent of that of the EU. Despite the image that exists of extensive air travel usage in the USA and within the EU, let alone between the USA and EU, it is Japanese domestic sectors which dominate the list of most used sectors (see Table 2.1). Why are so many people travelling so often in Japan? This is not a recent phenomenon as travel and tourism have a long history in Japan. Major roads linking across Japan have a history of over 1,300 years, although sea routes were also very important.[20] Whilst travel was restricted during much of the Tokugawa Period (1603–1858), permits could be received for religious travel.[21] This helped to give birth to early pilgrimages and such religious journeys were often a cover for other pleasurable exploits and 'partly because of these pilgrimages, the idea of travelling is considered very natural in contemporary Japan'.[22] Despite the advent of telephones and videoconferencing, the Japanese continue to put great importance on face-to-face contact. Many businessmen I have spoken to would not consider it unusual to do a 1,000 km round trip in one day in order to attend a one-hour meeting or to take a new year's gift to thank another company for their continued custom and business.

Table 2.1 Top air routes in the world

Rank	*Route*	*Region*	*Passengers in 2002 (million)*[1]
1	Haneda – Chitose	Japan Domestic	9.51
2	Haneda – Fukuoka	Japan Domestic	8.28
3	Haneda – Ōsaka[2]	Japan Domestic	7.32
4	Hong Kong – Taipei	Asia International	5.46
5	Seoul – Busan	Korea Domestic	5.39
6	Sydney – Melbourne	Australia Domestic	5.35
7	Seoul – Jeju Island	Korea Domestic	4.91
8	Haneda – Okinawa	Japan Domestic	4.33
9	London – New York	Europe – N. America International	3.90
10	Sydney – Brisbane	Australia Domestic	3.50

Table by the author.
1 Some data is from slightly different time periods.
2 Includes Itami and Kansai airports.
Generally transfer passengers are not included. Unless stated otherwise city names includes all airports within that city. Table based on JAL 2004. In 1985 the number of passengers on the Japanese sectors listed in the table were 4.56 m, 3.45 m, 3.10 m, 1.50 m respectively (JAL email 10-10-27).

In 1985, the expressway system was less developed than it is today and tolls had to be paid. In 1985 there were 3,600 km of expressways compared with 7,400 km in 2007 out of a total road network of around 1.2 million km.[23] Although the size of the expressway network has more than doubled, road transport is rarely a viable option for longer journeys, and it can be seen from Figure 2.1 that the modal share has dropped. From my experience of driving around Japan, part of the problem appears to stem from the lack of roads that perform a good inter-regional role. Local roads have many traffic lights and junctions, meaning that average speeds are low. Bypasses and routes which avoid city centres, as are found in Britain for example, are lacking. Dual carriageways are also lacking and overtaking on most roads is heavily restricted. Many of the expressways have only two lanes, and some only one. For example, the Bantan Renraku Dōro, which links Himeji and Wadayama in Hyōgo Prefecture is a toll road of some 64 km in length which links the industrialized southern part of the prefecture with the rural north, but for much of its distance the speed limit varies from just 60 km/h to 80 km/h and has only one lane in either direction for most of its length. Even on main expressways, where there are usually two, but sometimes three, lanes, speed limits are also relatively low (usually 100 km/h). The tolls have also been a disincentive to use these roads. While some have criticized the way in which Japan has continued to develop its expressway network,[24] for those who have driven around Japan, the feeling is more that the current network is merely providing the function of main 'A-roads' in Britain, and Japan is yet to develop a sufficient expressway network. Yet, there are still many who will use expressways, and at peak times of the year, such as around 12 August, it is not unusual for there to be traffic jams in excess of 30 km. Consequently, many others choose to travel by train or plane if that option is available.

Although Japanese trains tend not to have much space for luggage, the well-developed and cheap courier service means that most passengers are accustomed to sending larger luggage in advance. The alternative is to travel by plane. When considering the choice between train and plane, there are many factors which have to be taken into consideration, such as how to get to the station or airport. Within Tōkyō this can take more than an hour even if using public transport, and as many airports in Japan are not near city centres, the transfer time can be significant. There is then the waiting time at the station or airport, with significantly longer time usually needed for taking the plane due to the need to check-in and pass through security checks, for example. The next factor is the actual journey time itself. For a journey of around 500 km between Tōkyō and Ōsaka, the shinkansen takes 2 hours 25 minutes, compared with 1 hour by plane. However, even for this journey, when taking into account the other considerations, the shinkansen on balance is often the quickest option. Another factor that may need to be taken into consideration is the weather. Japan is regularly battered by typhoons, heavy rain and heavy snow. There are also the problems caused by earthquakes. When Mother Nature strikes, the news regularly covers the battle between the shinkansen and the planes as to which can best maintain its services.

As well as expanding the size of the expressway network and the shinkansen network, the LDP government also helped oversee the expansion in the many airports. In 1985 there were 78 airports across Japan, but by 2010 this figure had risen to 98.[25] Unfortunately the divisions within the Ministry of Transport have seemingly meant that a co-ordinated nationwide transport policy has rarely been considered.[26] The railway-*zoku* has sought the best deal for railways and the plane-*zoku* has sought the best deal for planes. Consequently no airport is served by the shinkansen as no consideration has been given to integrated transport in this way. Many of the new airports have been relatively small airports serving rural cities.[27] Agreements were put in place so that Japan's domestic carriers would serve these airports and it is arguable that such agreements were another of the factors which contributed to JAL's financial problems in recent years. From 1960 to 1986 there was growth in airline travel. The only years when numbers dropped were in fiscal years 1980, 1982 and 1985.[28]

The history of JAL until 12 August 1985

Japan Air Lines (*Nihon Kōkū*) was established on 1 August 1951 as a private company with a capital of ¥100m. JAL became the country's first post-war domestic carrier on 25 October that year when it began services of its Martin 202 plane, known as *Mokusei* (Jupiter), between Tōkyō, Ōsaka and Fukuoka.[29] However, *Mokusei* crashed on a flight from Tōkyō to Fukuoka via Ōsaka on 9 April 1952. All 37 passengers and crew were killed when the plane crashed on Mount Mihara on Ōshima island. In October 1952, JAL added Tōkyō–Sapporo, Tōkyō–Ōsaka and Tōkyō–Ōsaka–Iwakuni–Fukuoka routes using its new DC-4 plane, *Takachiho*. On 1 October 1953 JAL became a state-owned organization, assuming all the assets and liabilities of the private company, but with capital of ¥2bn.[30] On 2 February 1954 it began its international service with a flight, using a DC-6 known as *City of Tokyo*, between Tōkyō and San Francisco. The flight took 31 hours and involved stops at Wake Island and Honolulu.[31] On 12 August 1960, JAL acquired its first jet, a Douglas DC-8, called *Fuji*, which was used on the Tōkyō–San Francisco route.[32] In subsequent years JAL acquired other, larger capacity, versions of the DC-8. JAL added its first European route on 6 June 1961, using the polar route, with 'Silk Road routes' to Europe starting in October 1962.[33] At the same time JAL continued to upgrade its whole fleet to jets, having started introducing DC-8s in 1960 and using its first Boeing, a 727, on 1 August 1965.[34] By the end of the decade JAL had retired all of its non-jet planes (see Figure 2.2).

The change in the JAL fleet to jets was not merely about speed. The location of some of the airports in urban areas meant that the operation hours were relatively limited. Whilst Haneda was operational between 06:00 and 23:00, Itami was only operational between 07:00 and 21:00. Furthermore, the limitation on number of 'movements' (flights) was 420 and 370 respectively per day, but at Itami there was a further restriction of only 200 jet movements per day.[35] Whilst the DC-8 was

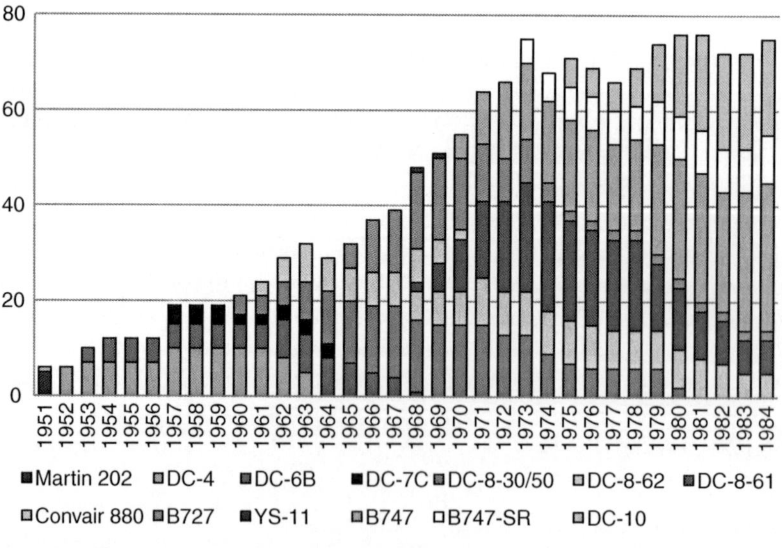

Figure 2.2 Changes in number of JAL passenger planes (1951–84)

Figure by the author. *Source:* JAL 2002:CD-ROM. *Note:* Data represents the number at the end of the fiscal year. Some aircraft we leased rather than owned.

Figure 2.3 JA8119
A photograph of JA8119 prior to 12 August 1985. Picture courtesy of JAL International.

becoming JAL's main plane, they were noisy and did not have sufficient capacity. With stricter environmental laws being introduced globally, the move towards an even larger, quieter and more economical plane was obvious. The 747 was the result. On 1 July 1970 JAL's first Boeing 747 entered service on its Pacific routes.[36] However, the typical 747 used on international routes did not offer the capacity which was sought on the Japanese domestic routes. Consequently the 747-SR (Short Range) was developed at the request of JAL.[37] Some of the key differences between the 747s are set out in Appendix 1. JAL took its first delivery of a 747-SR in 1973 (see Figure 2.2). Whereas the 727 had 129 seats and the DC-8-61 had up to 233, the first type of 747-SR had 498 seats.[38] Ironically, although Itami was the airport which had 'the most acute operating capacity problems', it was not served by the 747-SR until May 1977 due to a ban on wide-bodied operations relating to a 'protracted and complex law suit' seeking compensation due to aircraft noise.[39] JA8119 (see Figure 2.3) was introduced on 19 February 1974.[40]

Although JAL had managed to grow relatively successfully during the 1950s and 1960s, 'the domestic aviation system in Japan had seen a lot of turmoil', with a number of companies failing or merging. There was also an increase in international transport with many of the carriers looking to launch international services.[41] The government consequently looked to establish a new system relating to the operation of international flights. The system had been agreed by the Cabinet in Shōwa 45 (1970) and was fully implemented in Shōwa 47 (1972), and as a consequence is referred to as the '45/47 System' (*Yongo-Yonnana Taisei),*[42] or the 'Aviation Constitution of Japan'.[43] Under the system, JAL was given the national flag carrier status and was the sole Japanese carrier expected to provide international services, but could also operate on domestic trunk routes using Haneda/Narita, Itami, Fukuoka, Chitose and Naha/Okinawa. All Nipon Airways (ANA) was permitted to operate on domestic trunk routes and local routes as well as some short-distance international charter routes. Tōa Domestic Airlines (TDA) was allowed to operate on local domestic routes initially with the understanding that trunk routes would be allowed at a later date.[44] The reason for this policy was that 'the government did not want to see excessive competition' and 'saw air transport as an important source of tax revenue to fund the expansion of aviation facilities in Japan'.[45]

JAL continued to provide services for the Imperial Family, Japanese prime ministers and even the Pope until the Japanese government got its own Boeing 747-400s in 1991.[46] JAL was also not restricting itself to the skies. In 1975 it successfully tested its HSST, a Maglev train. This project was later shown off at the Tsukuba Science Exposition in 1985.[47] The concept behind the HSST was to provide faster links between city centres and airports. It was not expected to become the basis for inter-city transport, which has been developed using different Maglev techniques culminating in the Chūō Shinkansen test track in Yamanashi and the planned line between Tōkyō and Ōsaka.[48]

JAL grew from having around 800 employees in 1953 to over 20,000 in 1974. The level remained between 20,000 and 21,500 for the next ten years.[49] The number of flights had risen to some 63,000 in 1984, with about a 45:55 split between domestic and international flights.[50] Passenger revenues had also grown significantly (see Figure 2.4). Despite this, its financial performance fluctuated. Ordinary profits in 1984 were over ¥22bn, having risen significantly from ¥4bn the previous year and a loss of around ¥27bn in 1982 (see Figure 2.5).[51] Although the focus of this book is on passenger transportation, it is worth noting that JAL also saw comparable increases in freight transport, although the overwhelming majority was domestic.[52]

Despite the existence of the '45/47 System', the situation continued to change. First, due to the complications of the relationship between Japan, China and Taiwan, JAL's international services between Japan and Taiwan were suspended

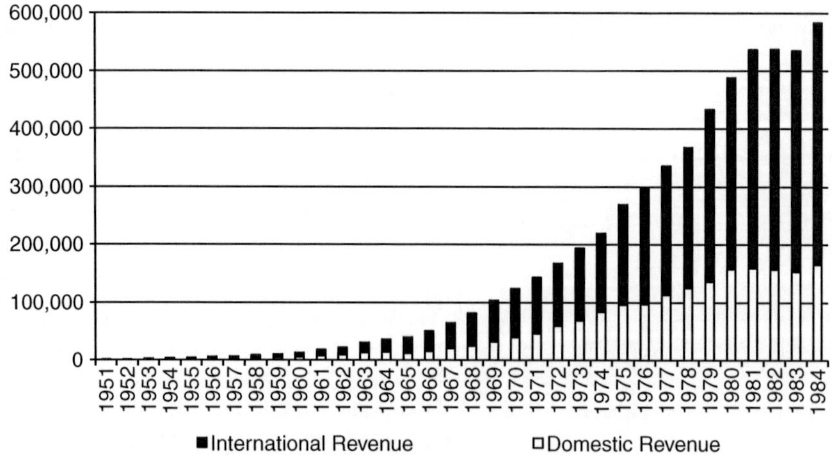

Figure 2.4 Changes in JAL revenues (¥million) (1951–84)

Figure by the author. *Source:* JAL 2002:CD-ROM.

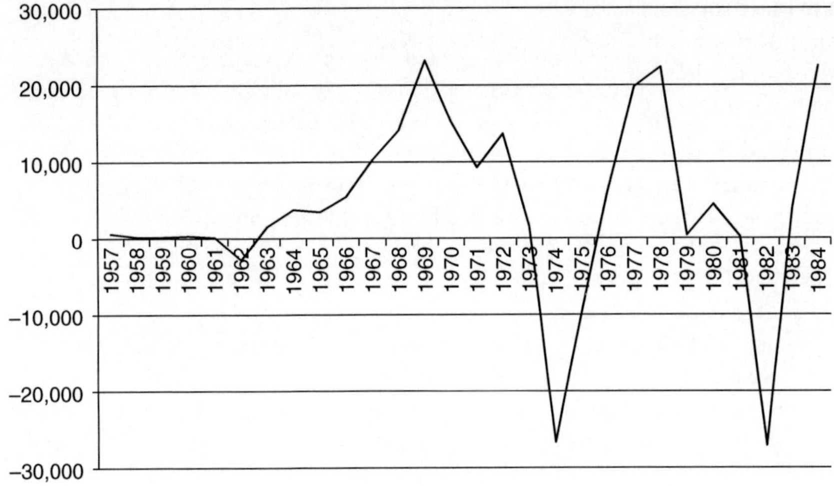

Figure 2.5 Changes in JAL ordinary profits (¥million) (1957–84)

Figure by the author. *Source:* JAL 2002:CD-ROM.

in April 1974, to be provided by a new company, Japan Asia Airways, from September 1975.[53] Naturally, ANA and TDA had hoped that this would have been their chance to have scheduled international services.[54] That this did not happen would appear to suggest a bias towards JAL within the government. Domestically, the three main companies had seen their relative strengths change. Between 1970 and 1984, JAL increased its assets by 3.6 times, compared with 8.5 for ANA and 6.5 for TDA. This was a significant point since Japan's domestic market had become the third largest in the world (after the USA and USSR) with more than 45 million passengers annually.[55] As a result there was 'growing consumer demand for more competition and better services'.[56] By the mid-1980s, the government was also looking to privatize a number of its assets. Other changes in the international aviation world and the creation of a new cargo airline by ANA and a consortium of shipping and freight companies 'finally triggered Japanese aviation policy change'.[57]

In April 1985 the government effectively announced the end of the '45/47 System'.[58] Within JAL there were moves to push towards its own privatization, being overseen by the president, Yasumoto Takagi. In 1981, Takagi became the first president of JAL who had worked his way up through the company rather than parachuting in from the Ministry of Transport in a process known as *amakudari*.[59] On 12 August 1985, a meeting was held at JAL headquarters where JAL's board of directors voted in favour of privatization to happen as soon as feasible.[60] After the meeting ended, the company president and others from the company attended a party in honour of a JAL-related company's president at the Takanawa Prince Hotel in Minato ward, Tōkyō,[61] unaware of the events which were starting to transpire which would greatly impact JAL's future.

Harumi Handa 半田春海 (56) · Michael Hanson (40) · Norihisa Harada 原田敬久 (28)

Airline disasters

1985 is a year that is remembered for aviation disasters. The JL123 crash came only a few weeks after another Boeing 747 was lost following a terrorist attack on an Air India jet killing all 329 aboard.[62] Ten days before JL123, a Delta Airlines Lockheed Tristar crashed on approach to Dallas-Fort Worth Airport, killing 137 people.[63] Ten days after JL123, a British Airtours Boeing 737 caught fire at Manchester Airport, leading to the death of 55 of those aboard.[64] On 12 December an Arrow Air Douglas DC-8 crashed on takeoff from Gandar, Canada, killing all 256 aboard.[65] 1985 was the worst year in terms of fatalities,[66] and we need to put these tragedies into context.

Statistics about aviation safety are numerous. For example, the number of commercial accidents is fewer than two accidents per one million departures worldwide.[67] According to one view, 'a baby born on an airliner, and staying airborne incessantly, would not be involved in a fatal accident for 82 years'.[68] Another estimate even suggests that you could 'take a flight each day for more than 30,000 years before expecting to perish in a plane crash'.[69] But accidents do happen and will continue to happen, partly due to the 'tombstone imperative' whereby companies, and government, will only make changes once enough people have died to make it necessary to make the change.[70] Although this view may seem extreme, the evidence is both compelling and depressing. It is also largely beyond the scope of this study, although relevant points will be introduced elsewhere.

Airline disasters are an area of fascination for many. They are among the most written and accessed stories, receiving 'media coverage disproportionate to their death toll'.[71] Books, documentaries and films that feature plane accidents are a popular genre. Yet, often the level of interest in them, and the level of concern that one may become a victim oneself, is out of proportion to the reality. Fear of flying tends to be one of the most common phobias in many developed societies, although the terminology is odd since surely it is the fear of the plane failing to fly, not that it is flying, which is the problem. Why do so many people have a fear of flying? Is it the perceived lack of control on a passenger jet that we are in the hands of an unknown, unseen pilot and the software provided by a computer? Is it that should something go wrong, we assume that death is probable? Although this is a common belief, around 43 per cent of accidents have survivors, which is about 1 per cent greater than the number of accidents with no survivors (data missing for the other 15 per cent).[72]

While it is largely, commercial, passenger-carrying plane accidents that get the attention, there are significantly more small plane accidents. In the 1990s in the US alone, 'general aviation' accidents account for around 2,000 accidents per year and an average of around 700 deaths per year. But 'unless the plane crashed into a school, church, or some other public venue, or involved a famous person, it is very unlikely that anyone outside the government or those intimately involved with the accident even knew it happened'.[73] So why is more attention and value put on certain lives and their deaths? This is a theme we will return to elsewhere in the book.

Most accidents occur during the 'critical eleven minutes' – the three minutes during and after take-off and the eight minutes of the landing phase. During this period the flight deck prefers not to be contacted by the cabin crew, except in an emergency.[74] According to ACRO (Aircraft Crashes Record Office), which keeps statistics for all aircraft accidents, some 21 per cent of accidents occur during take-off and 50 per cent during landing – a total of 71 per cent.[75] However, according to Boeing, looking only at commercial accidents between 1999 and 2008, 30 per cent of accidents take place before the plane reaches cruising altitude (a time which accounts for about 16 per cent of the total flight time), whilst a further 50 per cent occur during the descent and landing phase (which accounts for about 27 per cent of the flight time).[76] Accidents during the cruising phase of a flight only account for 8 per cent of the number of accidents, whilst it accounts for about 57 per cent of the flight time. ACRO similarly points out that only about 10 per cent of accidents happen in mountains, whilst nearly 54 per cent happen within 10 km of an airport.[77] In light of these statistics, JL123 appears unusual.

Figure 2.6 shows the top ten aviation disasters in the world and in Japan. As JL123 is the world's second largest disaster, and no other accident in Japan makes the top ten in the world, there are a total of 19 crashes listed. The China Airlines Flight CI140, the second largest in Japan, is currently the twelfth in the world. I have only listed those where at least one of the planes involved was a commercial passenger flight. If purely military flights are included, two would have made the world's top ten. Furthermore, how to deal with 9/11 is complicated. It is normal to include fatalities on the ground due to the plane crash as part of the total fatalities, but in the case of 9/11 the figures are so disproportionately high, as well as problems of associating deaths due to an individual plane (as is necessitated by there being four, albeit connected, incidents) in the case of the collapse of the World Trade Center. Consequently it is normal for the 9/11 deaths on the ground to not be included.

As can be seen from Figure 2.6, for accidents involving only one aircraft, JL123 has significantly more fatalities. Furthermore the JL123 and the Tenerife disaster are the only accidents to have over 400 fatalities. Looking at the domestic situation, JL123 saw double the number of fatalities as the second worst plane crash, which occurred nearly nine years later, and was over three times greater than the worst domestic aviation accident at that time, which was the world's worst aviation disaster when it occurred.[b] Although a number of the top accidents globally feature 747s, this should be seen as a reflection of the size of these planes rather than there necessarily being a safety issue with the plane itself. For example, although both of the planes involved in the Tenerife disaster were 747s, the design of the plane was in no way connected with the cause of the disaster. However, JL123 was the second 747 to be lost in 1985 and the possibility that it was the first not due to an act of terrorism or pilot error was one of the factors which drove the need for a rapid discovery for the cause, as will be discussed in Chapter 4. The 747 was seen as being so safe that the Japanese media referred to whether JL123 was ending the *shinwa* of the 747's safety record.[c] *Shinwa* is frequently translated as 'myth', but can also mean 'legend'. The nuance of these two words in English is quite different, and the latter would seem more appropriate here.

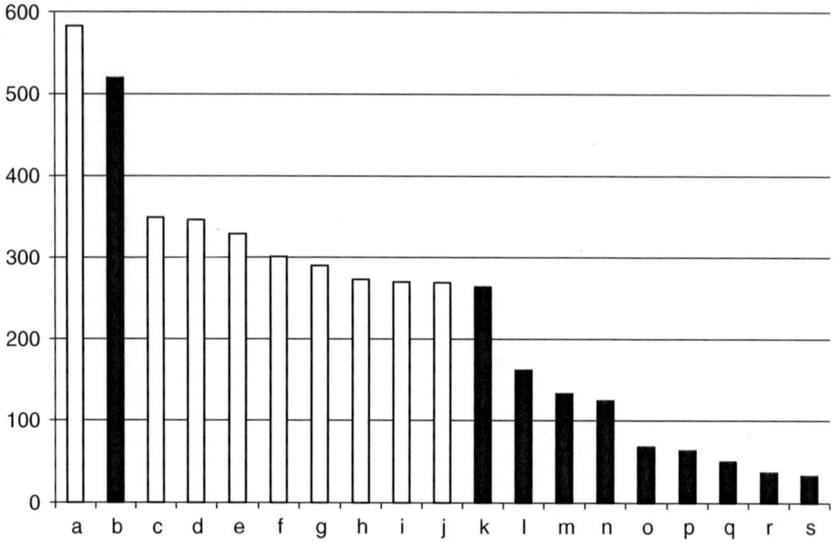

Figure 2.6 Top airline disasters globally and in Japan

Figure by the author. The figure shows the top ten airline disasters globally and the ten airline disasters in Japan (shown in black). The second worst accident within Japan is twelfth globally. Details correct as of 10/8/12. Deaths include passengers and those on the ground. Data only for civil airline crashes.

Sources: ACRO 2010c; ACRO 2010d; Asahi Shimbun 2003:218.

The details provided are Date of accident, Airline, Flight Code, type of aircraft and location of accident.

a: 1977-03-27 Pan Am PA1736, Boeing 747-121 and KLM KL4805, Boeing 747-206B, Tenerife, Spain; b: 1985-08-12 JAL JL123, Boeing 747SR, Osutaka-no-one; c: 1996-11-12 Saudi Arabian Airlines SV763, Boeing 747-168B and Kazakhstan Airlines K41907, Ilyushan IL-76, (midair) Charkhi Dadri, Haryana, India; d: 1974-03-03 Turkish Airlines TK981, McDonnell Douglas DC10-10, Ermenonville, France; e: 1985-06-23 Air India AI182, Boeing 747-236B, Atlantic Ocean near Ireland; f: 1980-08-19 Saudi Arabian Airlines SV163, Lockheed L1011-200 Tristar, Riyadh, Saudi Arabia; g: 1988-07-03 Iran Air IR655, Airbus A300B2-203, Persian Gulf; h: 1979-05-25 American Airlines AA191, McDonnell Douglas DC10-10, Chicago, USA; i: 1988-12-21 Pan Am PA103, Boeing 747-121, Lockerbie, Scotland; j: 1983-09-01 Korean Airlines KE007, Boeing 747-230B, West of Sakhalin Island, Russia; k: 1994-04-26 China Airlines CI140, Airbus A300, Nagoya Airport; l: 1971-07-30 ANA NH58, Boeing 727 and JSDF F-86 fighter jet, Shizukuishi; m: 1966-02-04 ANA NH60, Boeing 727, Haneda Offshore; n: 1966-03-05 BOAC BA911, Boeing 707, Mt Fuji; o: 1971-07-03 TDA JD533, Mt Yokotsudake (Hokkaidō); p: 1966-03-04 Canadian Pacific Airlines CP402, DC8, Haneda Airport; q: 1966-11-13, ANA NH533, YS11, Matsuyama Airport; r: 1952-04-09 JAL Mokuseigō, Martin 202, Mt Mihara (Ōshima, Tōkyō); s: 1958-08-12 ANA NH25, DC3, Shimoda Offshore.

Three of the Japanese aviation accidents happened in a four-week period in 1966, with two accidents on consecutive days. It is perhaps this that helped to cement the idea in some with whom I have discussed my research that accidents come in threes. Two of these accidents, although in Japan, involved foreign carriers. There have also been incidents involving Japanese airliners outside Japan, but within the context of this study, it is only pertinent to mention those involving JAL planes.

JAL has suffered a number of accidents throughout its history, but it has also suffered in other ways. Throughout the 1970s a total of eight JAL planes were hijacked. Although none of these hijackings led to the death of any passengers (other than hijackers) or crew, in one case in 1973, the plane, a Boeing 747, was blown up.[78] The 1970s also saw a number of accidents involving JAL. On 14 June 1972, a JAL DC-8 crashed near New Delhi's Palam International Airport, killing ten of the 11 crew and 72 of the 76 passengers, as well as a further three people on the ground. Although the Indian authorities believed the cause to be errors made by the crew, the Japanese authorities believed the cause to be an incorrect glide path being provided.[79] Later that same year, a JAL plane landed at an incorrect airport – Juhu Airport rather than Bombay Airport, as it was known at the time, about 5.5 km away.[80] Although nobody was injured, the runway at Juhu Aiport was not long enough for the DC-8 to take-off from and so the plane had to be taken apart. Barely two months later, on 27 November 1972, another JAL DC-8 crashed at Moskva-Sheremetyevo Airport when it stalled during take-off and burst into flames after hitting the ground. Nine of the 14 crew and 52 of the 62 passengers were killed in the accident.[81] For a number of years following this crash, JAL employees wore a remembrance badge of a green cross, a symbol of safety in Japan.[82] The next fatal accident occurred in Malaysia on 27 September 1977 when the plane struck a hill four miles short of the runway.[83] The next day JAL suffered yet another hijacking at the hands of the Japanese Red Army. The next fatal passenger-carrying service took place on 9 February 1982 when flight JL350 crashed on approach to Haneda Airport. Although none of the eight crew were killed, 24 of the 166 passengers were. The cause of the crash was found to be deliberate acts by the captain, who had been excused duty for a year before and who was seemingly trying to commit suicide.[84]

Yet 1985 was undoubtedly the worst year for JAL, the Japanese aviation world, and in many respects the aviation world as whole.

What is a disaster?

Although the term 'disaster' is used regularly in the media and in public discourse, we rarely stop to think about what the term is, as opposed to 'accident', 'incident', 'emergency' or 'tragedy'. Indeed, Levinson and Granot point out that 'there is no clear and universally accepted definition of when an accident becomes a *disaster*'.[85] It is often the response, or relative lack thereof, which can mean one event stops being merely an 'incident' and becomes a 'disaster'. Although it may be natural for academics to go searching for definitions and discuss them, this might not be the most appropriate way forward. As Cutter points out, sometimes too much energy is spent trying to define disasters, 'rather than researching more important and fundamental concerns' such as how we respond to the events themselves.[86] Consequently, I will just highlight some of the key issues relevant to this study so that more space can be devoted to discussing JL123. In terms of terminology relating to JL123, it is also worth stressing that there was both an 'accident' and a 'crash', but the two did not happen at the same time. As will be discussed in Chapter 3, the 'accident' took place around 12 minutes into the flight, whilst the 'crash' was the final moment of the plane's demise at around 18:56. However, many use 'accident' and 'crash' interchangeably to describe the whole event.

It is worth highlighting the two main categories of disasters: natural and man-made. Man-made disasters are sometimes referred to as technological, but this prevents the inclusion of acts of terrorism, for example. Natural disasters include earthquakes, typhoons, tsunami and so on, but there is a problem with the terminology, as it is not these natural events themselves which is the disaster, but rather society's response to them (including the degree to which society was prepared through the building of appropriate structures and response systems) which results in them *potentially* being a disaster. As Thang points out, the tragedy of the Great Hanshin Earthquake was the 'human error' by the government.[87] Part of the problem can stem from a lack of understanding of how we measure the size of earthquakes. An earthquake measuring 8.0 on the Richter Scale in an uninhabited area is unlikely to be considered a disaster, although its damage to the topography might be much greater than a 7.0 magnitude earthquake that strikes an urban area. Even in this case, whether the earthquake is a disaster or not is likely to vary from country to country, or even within a country depending on the level of destruction and loss of human life. Indeed, although it is common to refer to the Richter Scale, it is important to stress the limitations of this system. It is concerned with the intrinsic nature of the quake. Two 7.0 magnitude quakes can have totally different impacts depending on the geological and societal factors. To help take this into account, Japan usually uses its own system, the Japan Meteorological Agency Scale (JMS), similar to the Modified Mercalli Intensity Scale used in the United States, which measures the intensity of a quake. The Richter Scale is a logarithmic scale with each number meaning a factor of ten increase in amplitude.[88] The JMS ranges between 0 and 7 depending on the intensity at any given point.[d] In other words, for a single earthquake, different JMS figures will be produced depending on the intensity of the quake at different locations.

Japan is a country which is frequently hit by forces of nature. Although it is the large events such as earthquakes, typhoons and tsunami which grab the headlines and have accounted for the greatest loss of life from single incidents (see Table 2.2), it needs to be remembered that Japan faces a range of natural challenges. Sometimes people die trying to clear snow off the roof of their home so that the building does not buckle under the weight of snow. This can be a particular problem due to the traditional design of heavy roof which is supposed to help protect against typhoons. These roofs are also a problem during earthquakes as the top-heavy buildings are so susceptible to collapse. Japan also has four times the risk of landslides than most European and North American countries, which has been a particular problem for railway lines.[89] Similarly, whilst Table 2.2 has a number of entries for typhoons, generally the number of deaths attributable from a single typhoon is relatively low, whereas the number of deaths and damage in a year from several typhoons can be quite high. On average whilst around 27 typhoons are formed in a year, only 11 of these will approach Japan with just three striking land. In 2004 a record ten typhoons struck Japan,[90] leading to 220 fatalities.[91] In Japan typhoons are numbered according to their appearance in that calendar year, whereas the convention in many countries is to give the typhoon a name. A name has only been given to eight severe typhoons by the JMA (Japan Meteorological Agency), although many other typhoons have caused greater damage and loss of life.

Table 2.2 details a number of significant natural disasters that have struck Japan between the devastating Great Kantō Earthquake in 1923 and the Great East Japan earthquake in 2011 in chronological order. The events listed in the table are not necessarily the ones which have caused the greatest amount of damage in terms of loss of property or number of fatalities, but are ones which are either noted due to their size or impact in other ways. For example, the 2nd Miyakojima, one of the eight typhoons named by the JMA, resulted in no fatalities but had the fastest wind speed for a typhoon in Japan. The Chile earthquake of 1960 had no direct impact on Japan, but the tsunami created by the earthquake traversed the Pacific and resulted in 142 deaths in Japan. 2004, as noted above, was a particularly bad year for typhoons, with two of the ten that made landfall listed in Table 2.2, but there was also a significant earthquake that year too. One significant thing about this earthquake was the response of the government and the way in which the SDF was dispatched (see Chapter 9). The earthquake of 11 August 2009 is listed for two reasons. First, this was the second quake that I experienced during a one-week trip to Japan that year. I was in Takasaki at that time and woke shortly before the quake struck and then watched the news for the next few hours as the scale of the quake transpired. Second, earlier in 2009 I had seen an earthquake response prediction system being demonstrated at the National Research Institute of Fire and Disaster. Apparently this system quickly and correctly predicted that there would be only one fatality due to this quake.[92] To be spot on, given the size of the quake, strikes me as quite a significant achievement or an amazing coincidence. Although the number of fatalities per incident appears to have dropped over time, one cannot necessarily conclude that the lower death tolls are necessarily a reflection of better preparedness or responses by society as each event is so different. Each event is different, and the scale and nature of the earthquake and subsequent tsunami on 11 March 2011 showed that there is always potential for significant loss of life.

The focus of this study is on a man-made disaster. As a modern society, it should be of no surprise that Japan has suffered a number of large man-made disasters. Table 2.3 details some of the most significant man-made disasters in Japan since 1945 in order of the number of confirmed fatalities. This list includes the ten

Table 2.2 Some significant 'natural' disasters to have struck Japan (1923–2011)

Event	*Date*	*Earthquakes*		*Typhoon*		*Fatalities and missing*
		Magnitude	*Maximum intensity (JMS)*	*Lowest pressure (hPa)*	*Fastest wind (knot)*	
Great Kantō Earthquake	1923-9-1	7.9	Lower 6			142,807
Kita-Tajima Earthquake	1925-5-23	6.8	Lower 6			428
Kita-Tangō Earthquake	1927-3-7	7.3	6			2,925
Shōwa Sanriku Offshore Earthquake	1933-3-3	8.1	5			3,008
Muruto Typhoon	1934-9-21			N/A	N/A	3,036
Tottori Earthquake	1943-9-10	7.2	6			1,083
Tōnankai Earthquake	1944-12-7	7.9	6			1,223
Mikawa Earthquake	1945-1-13	6.8	5			2,306
Makurazaki Typhoon	1945-9-17			N/A	N/A	3,756
Nankai Earthquake	1946-12-21	8.0	5			1,330
Kasurīn Typhoon (No.9)	1947-9-15			N/A	N/A	1,930
Fukui Earthquake	1948-6-28	7.1	6			3,769
Tōya-maru Typhoon (No.15)	1954-9-21			956	75	1,761
Kanogawa Typhoon (No.22)	1958-9-21			877	175	1,269
Miyakojima Typhoon (No.14)	1959-9-12			905	165	99
Ise Bay Typhoon (No.15)	1959-9-21			895	165	5,098

Chile Earthquake Tsunami	1960-5-23	8.5	–			142
2nd Muruto Typhoon (No.18)	1961-9-8			888	185	202
2nd Miyakojima Typhoon (No.18)	1966-8-31			918	150	0
3rd Miyakojima Typhoon (No.16)	1968-9-17			930	120	11
Okinoerabu Typhoon (No.9)	1977-9-2			905	130	1
Central Sea of Japan Earthquake	1983-5-26	7.7	5			104
Mount Unzen volcanic eruption	1991-6-3					43
Southwest Hokkaidō Offshore Earthquake	1993-7-12	7.8	5			230
Great Hanshin Earthquake	1995-1-17	7.3	7			6,434
Typhoon No.18	2004-9-7			925	97	46
Typhoon No.23	2004-10-20			940	87	98
Niigata Chūetsu Earthquake	2004-10-23	6.8	7			68
Niigata Jōchūetsu Offshore Earthquake	2007-7-16	6.8	Upper 6			15
Iwate Nairiku Earthquake	2008-6-14	7.2	Upper 6			23
Suruga Bay Earthquake	2009-8-11	6.5	Lower 6			1
Great East Japan Earthquake	2011-3-11	9.0	7			27,897

Table by the author.

The date for typhoons is the date the typhoon first made landfall. After making landfall, typhoons typically continue to impact Japan for several days. N/A indicates that data was not available. Data for Great East Japan Earthquake as of 2011/4/17. *Sources:* JMA 2010a, JMA 2010b, JMA 2010c, JMA 2010f, JMA 2010g; Sugimoto and Nagai 2009.

Table 2.3 Some significant 'man-made' disasters in Japan (1945–2010)

Event	*Type*	*Date*	*Fatalities*
Methyl mercury poisoning/Minamata disease (Minamata)	Environmental	1932-1968	1,784
Tōya-maru rail-ferry sinking	Ferry	1954-9-26	c. 1,200*
JAL flight JL123 (Osutaka-no-One)	Aircraft	1985-8-12	520
China Airlines flight CI140 (Nagoya Airport)	Aircraft	1994-4-26	264
Hachikō Line Derailment	Railway	1947-2-25	184
Shiun-maru and *Washu-maru* sinking	Ferries	1955-5-11	168
ANA flight NH58 (Shizukuishi)	Aircraft	1971-7-30	162
Tōkaidō Line derailment and collision (Tsurumi)	Railway	1963-11-9	161
Jōban Line derailment and collision (Mikawashima)	Railway	1962-5-3	160
ANA flight NH60 (Haneda Offshore)	Aircraft	1966-2-4	133
BOAC flight BA911 (Mt Fuji)	Aircraft	1966-3-5	124
Fukuchiyama Line derailment (Amagasaki)	Railway	2005-4-25	107
Keihin Tōhoku Line fire (Sakuragichō)	Railway	1951-4-24	106
Hachikō Line collision (Tamagawa bridge)	Railway	1945-8-24	104**
TDA flight JD533 (Mt Yokotsudake)	Aircraft	1971-7-3	68
Canadian Pacific Airlines flight CP402 (Haneda Airport)	Aircraft	1966-3-4	64
ANA flight NH533 (Matsuyama Airport)	Aircraft	1966-11-13	50

Kintetsu Nara Line collision (Hanazone station)	Railway	1948-3-31	49
Shigaraki Kōgen Railway collision (Shigarakigūshi)	Railway	1991-5-14	42
Sangū Line derailment and collision (Rokken)	Railway	1956-10-15	40
JAL flight *Mokuseigō* (Ōshima)	Aircraft	1952-4-9	37
Meitetsu Seto Line derailment (Ōmori)	Railway	1948-1-5	34
ANA flight NH25 (Shimoda)	Aircraft	1958-8-12	33
Sarin gas attack on Tōkyō underground	Terrorism	1995-3-20	13
Mihama 3 nuclear accident	Environmental	2004-8-9	5
Tōkaimura nuclear accident	Environmental	1999-9-30	2

Table by the author.

* The exact number is not known as some bodies were not recovered.

** A further 20 bodies were not found. Sources: *Asahi Shimbun* 2003:218; Aoki *et al.* 2000:134–6; BBC 2006; *Nikkei Shimbun* 2010/3/7; Saigai Jōhō Sentā and Nichigai Asoshiētsu Henshūbu (eds.) 2007:28–41, 44–6, 52–6, 58–60, 132–8, 179–82.

aviation accidents already included in Figure 2.6. Additionally, it shows the ten largest railway accidents during the period as well as a number of other notable transportation disasters, environmental disasters and an act of terrorism. Due to its constitution, Japan is not permitted to be involved in warfare, so there is no need for this category to be included.

What is significant about many of the disasters in Tables 2.2 and 2.3 is the symbolism. Alexander suggests dividing the symbolism of disasters into three categories; '*functional* (i.e., symbolic of physical or social process), *linguistic* (i.e., a convenient form of notation) and *as an allegory or parable* (i.e., with a tale to tell, possibly of a moral kind)'.[93] The symbolism of disasters is one of the central themes of this study as it is an attempt to understand aspects of Japanese society by using JL123, and it is one which we will return to at appropriate points throughout the book.

If we take into account that a disaster is reflecting an inability for some reason, whether it be the sheer scale of the incident or society's lack of preparedness for example, to respond to a particular event, it is perhaps logical to suggest that a country used to dealing with events should be best suited to cope with them. If this is the case, then such a country should not have so many disasters. Japan *is* a country that has always had to deal with many natural hazards. Consequently, one would expect that Japan is accustomed to handling these and that a disaster after such an event may be due to the unexpectedly large scale of the event. By extension, one might also expect that Japan would also have the systems by which to handle man-made events. The possibility of such events is a reflection of the country's development. There is potential for traffic accidents, msanufacturing accidents and accidents in (nuclear) power plants. How these are responded to is likely to rely upon similar systems to those for natural events. However, I would argue that historically Japan has not responded well to either natural events or man-made events. The study of JL123 will demonstrate why this is the case and why it is appropriate to use the term disaster in this context.

Part II
The death of JL123

3 From take-off to disaster

> They say that it was thirty-two minutes from the 'bang' to the crash. It was a long time. It felt like hours. I waited for an announcement saying we were returning to Haneda. If that announcement came it meant that we could fly and we were communicating with the airport so we would be all right. But there was none.[1]

Having set the context in the previous chapter, it is now possible to discuss the details of flight JL123 itself. This chapter is provided as background to the main focus of the study which follows in the subsequent chapters. This chapter does not cover every detail of the crash; some parts will be discussed in more detail in other chapters. However, there are other areas which are either not possible to include, or I have decided not to include. For example, one of the survivors, Yumi Ochiai, provided a very detailed and moving account of her memories of the flight, the crash and its aftermath. This was done through interviews with Shinobu Yoshioka and subsequent checking of the text over some time in 1985, and updated the 'Ochiai Statement' given on 14 August which contained a number of problematic areas (discussed elsewhere). The result, a part of which is above, was then included in Yoshioka's award-winning book *Tsuiraku no Natsu*,[2] which won the Kōdansha prize for non-fiction books in 1987.[3] A translation of most of the interviews can be found in Weir's excellent book *The Tombstone Imperative*.[4] Although providing all of this text would undoubtedly add much to this chapter, it would reduce the amount of text and analysis in subsequent chapters due to length restrictions of the book as a whole. I also considered including the full text of the CVR, but did not due to both space limitations and the concerns over the accuracy of the transcription.

As mentioned in Chapter 1, the lack of English studies of JL123 was one of the motivating factors for me to write this book. The only book to contain a description of the JL123 crash comparable in length to this chapter is Job's chapter 'JL123 – Uncontrollable!' in *Air Disaster Volume 2*, although that one contains the discussion of the cause of the crash which is covered elsewhere in this book.[5] As discussed elsewhere, while the amount of printed material in English relating to JL123 is limited, there is no shortage of material in Japanese and this chapter uses information from a variety of sources, although using details from the official report as the main

Toshiko Higashitani 東谷敏子 (36) · Akiko Higashitani 東谷亜紀子 (10)

basis. When citing from the official report, I have generally provided the English reference details, given that this book is primarily for an English-speaking audience. However, as I have concern about the translation of the CVR, I have used the Japanese original as the basis for the text in this chapter.

I had not originally planned to discuss the cause of the crash itself in this research. My original intention was that this chapter would present the events as they happened in a chronological order. This in itself would be a break from the standard conventions of aircraft accident writing and documentaries, whereby the flight is usually presented in chronological order without necessarily clarifying the reason for the accident, as this would be done – as if through the eyes of the official accident investigation team – subsequently. However, after reading many of the Japanese sources I realized that a discussion about the cause of the crash is necessary and this is done in the next and subsequent chapters, with this chapter focusing upon the facts, as they are known, about the flight itself.

Prior to take-off

The plane that crashed had the registration number JA8119 and was delivered to JAL on 19 February 1974.[6] It was a Boeing 747 SR-46.[a] To the layperson, it was a 'jumbo jet'. However, it was specially adapted to cope with the strains of additional take-offs and landings due to it being used on domestic routes (see Chapter 2). The plane was painted in JAL's traditional livery of being predominantly white with a blue and red stripe along the body. On the tail was JAL's famous logo of a red crane (see Figure 2.3, page 25).

Prior to 1985 JA8119 had been involved in two other accidents during its 11 years of service. The first of these was at about 15:00 on 2 June 1978 at Itami, when the plane landed with its nose up at too much of an angle, leading to the plane striking its tail on the runway (see Figure 3.1). Of the 375 passengers on board, two had to be taken to hospital suffering from shock.[7] Part of the rear bulkhead, which separates the passenger area and has to be pressurized at high altitude from the unpressurized tail section, was damaged in the accident and had to be repaired by Boeing engineers who flew to Japan (see Chapter 4). JA8119's second accident was on 18 August 1982 at Chitose Airport. In this case the damage, again occurring during landing, was to the number four engine (the outer engine on the right wing) when it struck the runway.[8]

Table 3.1 shows the schedule for JA8119 on 12 August 1985 and reveals the great usage that jumbo jets had on the domestic routes. JA8119 had been flown for a total of 25,030 hours and 18 minutes and made 18,835 landings,[9] which equates to an average of 1 hour 20 minutes per flight. JA8119 arrived at Haneda Airport as flight JL366 from Fukuoka at 17:12, slightly behind schedule, and five minutes later parked in spot 18.[10] Other than the flight engineer, who had also been on flights JL363 and JL366, the cockpit crew was different for JL123.[11] The cockpit crew comprised of a captain, co-pilot and flight engineer. The captain would usually sit in the left-hand seat, but for flight JL123 the experienced captain was sitting in the right-hand seat so that the co-pilot could fly the plane

Figure 3.1 JA8119's 1978 accident

Photograph courtesy of *Mainichi Shimbun*.

to gain additional experience as part of his on-the-job training. Captain Masami Takahama was 49 years old and had joined JAL in December 1966. He obtained his pilot's licence in 1969 and was permitted to fly 747s from 1975. In 1977 he qualified as a captain who could authorize his co-pilot to fly the aircraft from the left-hand seat. He had a total flying experience of 12,423 hours and 41 minutes, of which 4,842 hours and 22 minutes were in 747s. He had last received flight simulator training in June 1985 and also had other training earlier that year. In the previous 30 days he had flown for 53 hours and 46 minutes.[12] The co-pilot, Yutaka Sasaki, who would be flying the plane, was 39 years old. He had joined JAL in April 1970 and had got his full pilot's licence in 1984. He had qualified to be a left-seated co-pilot in 1984. He had a total flying experience of 3,963 hours and 34 minutes, of which 2,665 hours and 30 minutes were in 747s. He had last received flight simulator training in May 1985 and had also had other training earlier that year. In the previous 30 days he had flown for 46 hours and 47 minutes.[13] The flight engineer, Hiroshi Fukuda, was 46 years old and had joined JAL

Yasushi Honda 本田寧 *(38) · Haruyoshi Hongō* 本郷晴喜 *(41) · Tatsuo Horioka* 堀岡竜夫 *(53)*

Table 3.1 JA8119's schedule for 12 August 1985

	Flight No.	*Origin*	*Departure Time*	*Destination*	*Arrival Time*
1	JL503	Haneda	07:55	Chitose	09:20
2	JL504	Chitose	10:10	Haneda	11:50
3	JL363	Haneda	12:55	Fukuoka	14:35
4	JL366	Fukuoka	15:30	Haneda	17:00
5	JL123	Haneda	18:00	Itami	19:00
6	JL130	Itami	20:40	Haneda	21:40

Table by the author.
Sources: AAIC 1987b:6; Aoyama 2010:129; JAL interview.

in April 1957 and had got his flight engineer's licence in 1965, and the licence for a 747 in 1972. He had a total flying experience of 9,831 hours and 3 minutes, of which 3,846 hours and 31 minutes were in 747s. He had last received flight simulator training in February 1985 and had other training within the previous 12 months. In the previous 30 days he had flown for 41 hours and 26 minutes.[14]

Passengers were served by a total of 12 pursers, including one male chief purser, all of whom had remained on the plane from the previous flight.[15] The chief purser was 39 years old and had a total of 10,225 hours and 33 minutes flying experience. The other 11 pursers[b] were all female, aged between 24 and 31 and each had flying experience ranging from 363 hours and 41 minutes to 5,704 hours and 55 minutes. They had all received rescue training within the previous 15 months, with half having received training on the same day in June.[16] The chief purser was seated by exit door L1 (see Figure 3.4), one purser was in the first-class upper deck, and the other pursers were seated next to each of the emergency doors, with two being next to R4.[17]

Being 12 August, many people were leaving Tōkyō in order to return to their ancestral homes for Obon (see Chapter 6). As a result of the mass migration which had taken place during the 1960s in particular, many Japanese live far away from the ancestral home but continue to return there to take part in the traditions of visiting relatives and the ancestral grave at the time of Obon. On top of this, it was the school and university holidays, when many combine travelling to Obon with taking a holiday, particularly as in the 1980s there were many who still felt a 'need to justify personal enjoyment'[18] at a time when working hours were among the longest in the world and many Japanese did not take holidays.[19] This time in August consequently becomes the busiest time in the calendar for inter-city and international travel. Figure 3.2 provides a summary of the types of passengers that were aboard JL123. The average age of the passengers was 34.4 years (compared with 31.7 for the crew), although the average age was higher for male, who made up 68 per cent of the passengers (see Figure 8.2, page 180). Of the 520 passengers, 92 were aged 19 or younger,[c] with 49 being younger than ten.

Kōichi Horiuchi 堀内幸一 (54) · Junji Hosokawa 細川順治 (54)

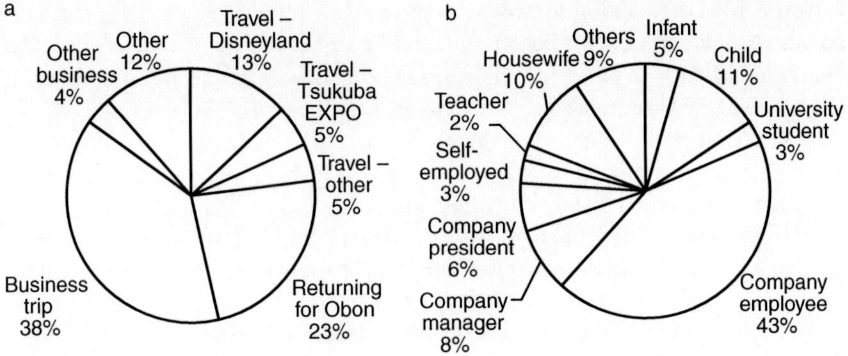

Figure 3.2 Types of passengers on JL123

Figure by the author. Data from 8/12 Renrakukai 1987:294.

As mentioned in the Miyazaki quotation in Chapter 2, JL123 had a large number of people from the Kansai area. Figure 3.3 shows the regional backgrounds for the passengers on board. I have come across cases where local newspapers subsequently claimed that a passenger was from their area and a prefecture other than the one listed in the passenger list in the national newspapers, for example (see Chapter 7). However, the number of such variations is probably very small. One can see that Ōsaka and Hyōgo prefectures alone accounted for 299 of the 524 (57 per cent) on board.

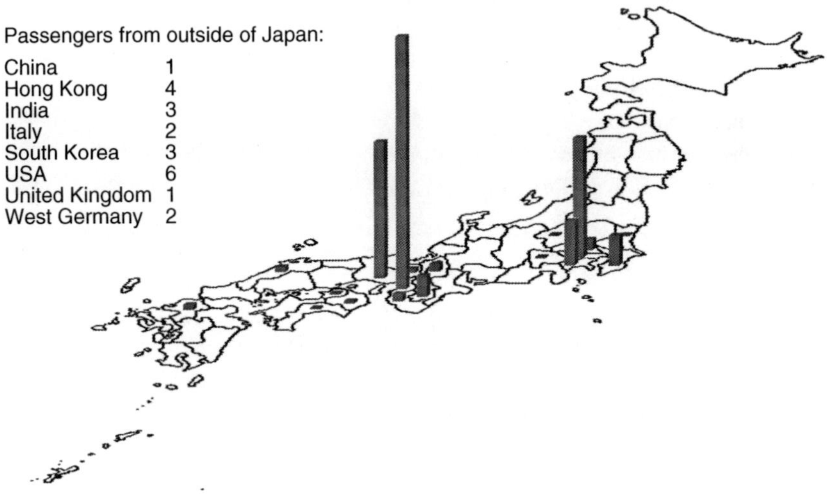

Figure 3.3 Regional background of passengers and crew

Figure by the author. Data based on 8/12 Renrakukai 1987:294 and newspaper articles. Note that some of the foreigners were resident in Japan. There were also some who had dual nationality.

Kazuhiko Ichikawa 市川一彦 *(53)* · *Akira Ichiura* 市浦昭 *(47)* · *Umeji Ide* 出梅治 *(33)*

It was a warm evening in Tōkyō as passengers boarded JL123 with the temperature still around 29 °C after having reached a peak of 31.5 °C that day,[20] whilst the humidity was 72 per cent.[21] There was no gangway connecting the plane to the terminal, so passengers had to cross a patch of tarmac before climbing the steps up to the plane. Some passengers stopped to take photographs before boarding. Inside the terminal some passengers rushed to catch the plane. Some made it. Others were turned away and would have to take another flight later that day. Although there were some spare seats available (see Figure 3.4), the plane was nearly full when the number of infants (sitting on an adult's lap) was taken into consideration.[d] In total there were 509 passengers on board.

Although Figure 3.4 details the seating plan based on the passenger manifest, it is uncertain whether passengers were located in these seats at the time of the crash. It is for this reason that I have not identified the passengers in this diagram or provided the seating details in the passenger list in this book. There are two particular cases which lead me to question the seating plan. First, I have seen a photograph taken inside JL123 by one of the passengers, Tetsu Ogawa, prior to take off. This picture shows his wife and daughter sitting next to each other with his daughter in the window seat. However the seating plan has Mrs Ogawa in the window seat (56K) and Mr Ogawa next to her (56H) with their daughter sitting in a seat in the middle section of the plane (56G). Further photographs, probably taken by Mr Ogawa (see Chapter 8), would also suggest that he was seated in 56G. In another case, the passenger manifest suggests that a mother was sitting with her small child on her lap, but seven rows away from her seven-year-old daughter. This is despite the fact that there were places where two vacant seats were located next to each other. One is left wondering whether these three were indeed separated during the flight, and the torment that this must have caused as the tragedy unfolded, or whether they were in fact seated together after they boarded the plane and the availability of adjacent seats became clear.

After the pursers helped passengers to their seats, the doors were closed and JL123 made its final departure preparations.

Take-off

JL123 officially departed Haneda 'on-time' at 18:04,[e] and a few minutes later JL123 lined up on Runway 15L. After receiving clearance from the control tower, it accelerated along the runway and took to the skies, taking off at 18:12.[22] On the ground, a plane enthusiast video recorded the take-off and watched as JL123 climbed away from the airport.[23]

The scheduled route would take JL123 via Mihara, Sagara and the Kushimoto VORTAC beacon. The scheduled flight time was 54 minutes and the plane had enough fuel for 3 hours and 15 minutes of flying.[24] At 18:16:55 the pilot contacted the Tōkyō area control centre (ACC) and asked for permission to change to another vector which would allow the plane to catch up some of the time lost during the delayed take-off. Permission was granted at 18:18:33.[25]

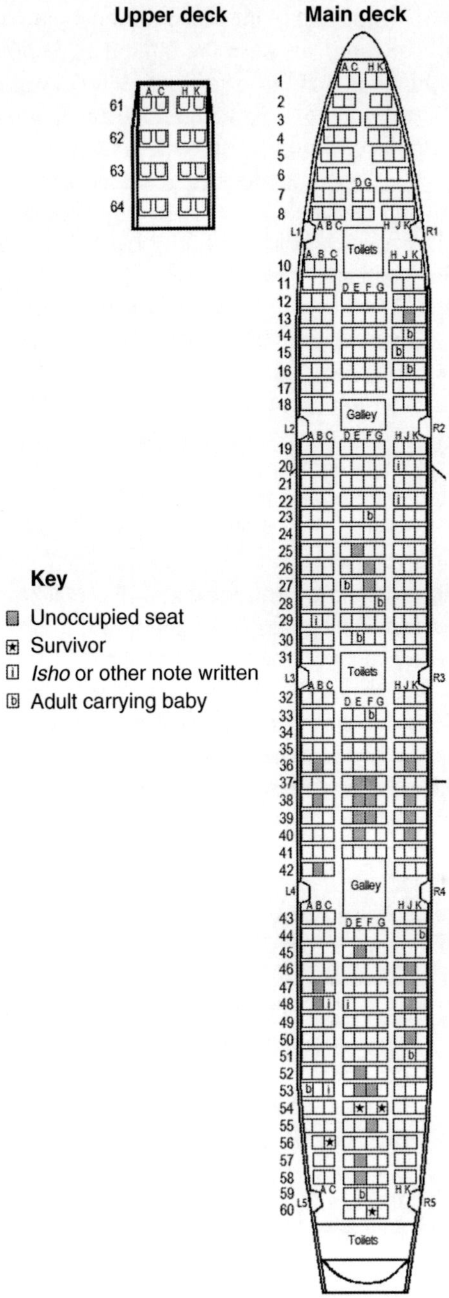

Figure 3.4 JL123 seating plan

Figure by the author. *Notes:* Figure is for illustrative purposes and is not to scale. Passengers may not have sat according to their seat allocation (see main text). The purser who wrote an *isho* was sat near L5.

What else happened inside and to the plane during the next few minutes until 18:24:35 is not totally known. Data from the flight data recorder (FDR) does not appear to reveal anything anomalous. The FDR records data relating to various aspects of the plane's performance, speed, direction and the forces acting upon it. The FDR, together with the cockpit voice recorder (CVR), are commonly referred to as the 'black box'. Black boxes are located at the rear of the plane and are actually orange to make them easier to find should they need to be recovered from dark locations (e.g. underwater). In the case of JL123, both the FDR and CVR were recovered at the crash site. The CVR records all sounds in the cockpit on a 30-minute-loop tape. In other words, after 30 minutes the machine begins to overwrite the material that occurred 30 minutes earlier. A 30-minute loop is the norm, but there are some tapes longer. JA8119's tape was about 32 minutes long. By chance, this meant that the events at 18:24 remained. Had the plane crashed just 23 seconds later, this evidence would have been lost.

The CVR is set up to pick up sounds and conversations from different sources. There is an overhead microphone within the cockpit which records general sounds and conversations. A second channel – linked to the microphone of the person in the right-hand seat on the flight deck – picks up communication with ACC and conversations to and from the co-pilot. A further channel – linked to the microphone of the person in the left-hand seat on the flight deck – picks up the communication over the loudspeakers in the cabin of the plane. Naturally, the CVR overhead microphone also records the pilot's or co-pilot's words if they operate their own microphones. In addition to the CVR, there are other records of conversations between JL123 and the ground and within JL123. First, there were the communications with ATC facilities. This largely relates to the preparation and departure of the plane and covers a time period from 17:53 to 18:18. Second, there was a record of the plane's communication via the company phone to JAL. This only recorded two short conversations – one prior to take-off and one a few minutes after take-off. It would appear that at this point, at 18:21, there was no cause for concern.

The CVR recording begins from 18:24:12 picking up the second part of a conversation between a stewardess, the co-pilot and the flight engineer. 'There is a person who would like to X. Is that OK?' Due to the nature of the Japanese language, 'X' came on the part of the tape which was written over later on in the flight. It is assumed that the purser was saying that a passenger was asking to go to the toilet. The significance of this question is that it means that the seat belt sign must have been turned on at this point (see Chapter 4). The captain stated that they should take care, as did the flight engineer. The captain added 'be quick'. The stewardess thanked them, to which the flight engineer once again asked her to take care. The recorded part of the conversation took only about six seconds.[26]

Disaster strikes

At 18:24:35, just as the plane reached its cruising height of 24,000 ft,[27] 'an explosion' can be heard. Japanese often uses onomatopoeia, but despite this there can be variations on how sounds are described. The AAIC refers to the sound as being a '*dōn*',[28] or 'boom' or 'bang' in English.[29] Survivor Yumi Ochiai, who was sitting near where the explosion occurred, described the sound as being a '*pān*'.[30] In fact, there was more than one sound, as noted by the report itself and others such as Yoneda.[31] Consequently, Yoneda describes the sound as '*dodōdodon*' and '*dodōdon*'.[32] Whatever the sound(s), clearly something significant had occurred. A fuller discussion about what may have happened will follow later. With the explosion, much of the rear stabilizer (tail fin) was lost. Other parts of the tail were lost at this time and during the remainder of the flight. However, with no sensors in that area of the plane, no direct view from any part of the plane and no camera filming that section of the plane, it would appear that the full extent of the damage was never known to anyone onboard JL123.

Barely a second after the explosion the captain suggested that something had exploded and two seconds later declared an emergency by signalling 'Squawk 7700', the transponder code for a general emergency.[33] At the same time there appeared to be some concern about the landing gear doors and possibly something to do with the engine. However not everyone, for example Yoneda,[34] agree about the contents of the official transcript for this part of the CVR. In the passenger section the chief purser was announcing to all the passengers that they should put on their oxygen masks, which had dropped down automatically. He repeated the request twice and asked people to put on their seat belts.[35] At 18:25:14 the pre-recorded announcement (PRA) started; asking passengers to fasten their seatbelt, put out their cigarettes and stating that the plane was in an emergency descent. Both Japanese and English announcements were made.[36]

Thanks to a photograph probably taken by Tetsu Ogawa (see Figure 3.5), it is known that the oxygen masks dropped in the cabin section and that pursers, donning portable oxygen masks, were helping passengers with their own masks. One has to be careful about drawing too many conclusions about the state of passengers and their behaviour from a still photograph. The stillness projects calm. That no person making any sudden or obvious movement is captured in the photograph helps to support this. Although Yumi Ochiai said that there was no panicking at first,[37] it is unlikely that the situation was as calm as this photograph suggests. Although it is unclear what time exactly the picture was taken, it is likely to have been a few minutes after the initial explosion – at a time when we know some passengers were writing *isho* (see Chapter 5). The fear revealed in some of these, is perhaps also seen in the photograph where one man appears to be gripping the seat in front of him.

Meanwhile in the cockpit, the co-pilot made a comment about the hydraulic pressure. At the same time as much of the rear stabilizer broke off, the hydraulic pipes had been severed. With no valve in place to prevent leakage, all of the plane's hydraulics would be lost. This meant that the lack of rear stabilizer would have been less obvious since control had been lost to all the moveable parts within the stabilizer.

At 18:25:21 the captain contacted ACC, 'request from immediate… trouble request return back to Haneda descend and maintain 220 over.' In other words, he was confirming the 'Squawk 77' emergency signal and was requesting an initial

Figure 3.5 Inside JL123

Photograph courtesy of Ryōichi Ogawa.

descent to 22,000 feet and for a route back to Haneda airport. The request was immediately accepted. That the request was for a descent to 22,000 feet is significant and will be discussed in Chapter 4. The captain also asked for a vector to turn to Ōshima. After checking whether he wanted to make a left or right turn, ACC confirmed a '090 radar vector to Ōshima'.[38] But instead of turning back towards Ōshima, the plane only turned slightly and largely proceeded in a westerly direction. During the next minute, the three men in the cockpit confirmed that all hydraulics had been lost. At 18:27:00 ACC contacted JL123, 'Japan Air 123 confirm you are declare emergency that's right?' 'That's right', the captain replied. '123 roger. And request your nature of emergency.' There was no response to this question. Perhaps the captain was distracted by another warning buzzer within the cockpit which sounded at about this time. Fifteen seconds later the captain asked about the state of the hydraulics. After acknowledging the question, the flight engineer took about 15 seconds until he confirmed that all hydraulics had been lost.[39]

The co-pilot then requested that they contact JAL on the company phone. The captain appears to have replied by asking 'Why make a fuss?' This question would seem slightly puzzling. Does it suggest that the captain, although aware that they were in a difficult situation, did not yet consider it to be one that could not be handled? Whatever his reasoning, no communication was made directly to JAL at this time. A further 14 seconds elapsed before the flight engineer suggested that they should descend. A further 15 seconds later, JL123 was contacted by ACC, 'Japan Air 124, fly heading 090 radar vector to Ōshima'. This command was further confirmation of the request of position made by the captain three minutes earlier. However, ACC identified the plane with the wrong flight number. Despite this, the captain of JL123 replied with the chilling words 'But now uncontrol'.[40]

By this time the plane had entered into a phugoid motion as well as a 'Dutch roll'. A phugoid motion,[f] also known as 'porpoising', is caused by a plane losing altitude when it has insufficient speed to maintain its current altitude. However, as the plane descends, it naturally gains speed and in doing so its pitch increases until it begins to increase its altitude until a point whereby the cycle is repeated.[41] 'Dutch roll' refers to a rolling and yawing motion and is derived from the motion made by skaters along Dutch canals whereby the body rotates from side to side.[42] These two motions acting together would have made the experience inside the plane particularly unpleasant. Although Yumi Ochiai described it as relatively slow, rather than sudden, movements or vibrations at first, she says that the motions got greater later on to the extent that she could not stand.[43] Whilst a roller coaster tends to go up and down or bank from side to side, they rarely do both at the same time. JL123 did. Whereas with a roller coaster one knows that there is a limited time until the ride should come to a safe end, for those on JL123 this reassurance was lacking.

The pilot continued fighting with the plane. At 18:30 the captain asked what something was, but what he was looking at and who the comment was directed to is unknown. Thirty seconds later the flight engineer was engaged in a conversation on the intercom with one of the pursers. There is only a record of the flight engineer's participation in the conversation. He first asks about the oxygen pressure and then whether the oxygen masks have dropped. After a pause – during which time he presumably got a response – he continues to say something about the oxygen pressure and using portable oxygen bottles. Exactly what was said is not totally clear. However, about ten seconds later, perhaps after getting some further information from the person he had been speaking to, he tells the others in the cockpit that the oxygen masks had dropped. He confirms this again about ten seconds later, by which time the captain is engaged in a further communication with ACC. At 18:31:00 ACC contacted JL123, asking 'can you descend?' The captain confirms that they are doing this. ACC asks for confirmation of the altitude, to which the captain replies '240'. In other words the plane was now at 24,000 feet – 2,000 feet above where they had requested to descend to some six minutes previously, and at the same altitude as when the problems began. But during this six-minute period the plane had gained and lost altitude four times, getting down to around 22,000 feet at one point (see Figure 4.8).[44] By now the plane was about over Yaizu, Shizuoka prefecture (see the route map, page xix). ACC asked whether they wanted to continue to Nagoya, which was '72 miles' (116 km) away. The captain rejected this and stated his preference to return to Haneda,[45] now approximately 95 miles (150 km) away. We can only speculate about the motivation to make the return to Haneda. Given that Nagoya Airport is located in a heavily populated area, compared with Haneda's location on a man-made island, one can assume that the pilot was concerned about the consequences of not landing on a runway. It is also likely that the pilot had more experience of

Haneda than of Nagoya Airport, and so had greater familiarity with the landing procedures and so forth for Haneda.

Of course there is no way of knowing exactly what the captain was thinking or what his intentions were, as these were never discussed with the cockpit crew, let alone with ACC. One has to assume that they were hoping to stabilize the problem with the plane, to get it under control, and then to make an emergency landing at Haneda. There was clearly an awareness of a problem, but the rapidness to declare an emergency could have been as much an automatic reaction to the sound of the explosion as being a realization that there was a serious problem with the plane. This latter point appears to be supported by the captain's reaction to the co-pilot's request to contact JAL headquarters. Although hydraulics were being lost, which was seemingly greeted with disbelief, there may have been a hope at this stage that the situation was recoverable. Consequently, considering an emergency landing on water would have been highly unlikely, and, by the time this may have become a consideration, the plane was uncontrollable and far from the sea. In fact, until the landing of US1549 on the Hudson River in 2009, a successful emergency landing on water by a commercial passenger jet was largely unheard of. Indeed, the Aviation Consumer Action Group suggested that a wide-bodied jet landing on water would 'shatter like a raw egg dropped on pavement, killing most if not all passengers on impact, even in calm seas with well-trained pilots and good landing trajectories'.[46] Although passengers receive information about such procedures and where to find their life jackets, which 'have little purpose other than to make passengers feel better',[47] the fact is that planes do not like landing on water. The assessment of the Hudson River was that the pilot had 'succeeded in doing something *unique*' (emphasis added) and there was a degree of luck.[48] Comparisons were made with the Ethiopian Airlines flight ET961 crash in 1996, but as Bartlett points out, the conditions were very different.[49] In the case of ET961, the sea had a swell rather than being on a smooth river, the pilot was having to fight off a hijacker, and the plane had no auxiliary power unit (APU) to generate power as the fuel had run out.[50] The result was that the plane's left wing hit the water first, sending the plane into a spin, causing it to brake up on impact. It is a miracle that 50 of the 175 on board survived, and was helped by the plane being relatively close to a beach – although more people might have survived had they followed to the captain's instructions to not inflate their life jackets before leaving the aircraft.[51] In the case of JL123, although there was no hijacker, there was no APU (this came off around the time of the explosion) and the sea would not have been smooth. The chances of keeping the plane level enough to make an emergency landing would have been remote and given the situation and the low probability for a successful landing of this kind, I think it is totally understandable that captain Takahama did not attempt to land on water.

Immediately after the captain said that he wanted to return to Haneda rather than go to Nagoya, ACC said, in Japanese, that all further communication could be done in Japanese. This was confirmed by the captain.[52] It might seem surprising that, even for a domestic flight, communication would ordinarily take place in English, but one has to remember the open nature of the air. The communication between ACC and JL123 could be, and was being, monitored by others. If these planes were unable to understand each other, then it increases risk. However, the form of English used in the communications between JL123 and ACC would suggest that a stylized form of English, using some aspects of Japanese linguistics and grammar, has evolved.

Trying to regain control

Exactly seven minutes had passed since the explosion, and the flight engineer appeared to engage in a further conversation with one of the pursers. It is not clear whether this was with the same person as before or not as again only the flight engineer's involvement in the conversation was recorded. Contact was initiated by the purser and the flight engineer's first comment was, 'Yes, what is it?' Given the fact that there was an on-going emergency, one has to assume that the nature of the information was very important as pursers are not expected to contact the flight deck during critical times (including the 'critical eleven minutes' (see Chapter 2)).[53] After a further pause, during which time the purser gave their information, the flight engineer asked, 'Towards the back?' The following very brief pause would suggest that the answer to that question was 'yes'. The flight engineer then appeared to ask if something was broken, although the CVR is not totally clear. After another short pause, although longer than the previous one, the flight engineer then asked, 'Where is it?' The answer to this question took about four seconds. The flight engineer then spends some time apparently confirming this location with the purser, before repeating the information to the captain and co-pilot: the location of the damage is apparently at the back of the plane and the overhead baggage hold has collapsed. He suggests that they should land and then adds after about ten seconds that everyone is wearing their oxygen masks.[54]

About 45 seconds later, at around 18:33:15, it would seem that the flight engineer was contacted again. What was said to him is not recorded, and his reply is not clear. It would appear that it was something to do with oxygen. After his response, there is a further pause before he speaks again. In his next conversation he mentions 'R5', but the whole sentence is not clear. 'R5' refers to the rear-most door on the right-hand side of the plane (see Figure 3.4). After that he says, 'Yes, understood' three times with brief pauses in between, suggesting that additional information is being relayed to him. He then relays the information to the captain. Again, there is some debate about what is said, but it is suggested that it may have been something to do with the oxygen masks near R5. The flight engineer then suggests that they make an emergency descent, which the captain confirms.[55]

Two seconds later, at 18:33:45, the flight engineer then suggests that the cockpit crew put on their oxygen masks. The captain says 'Yes' and the co-pilot also says that it would be best if they put them on. The flight engineer again suggested that the oxygen masks be used, and again the captain said 'Yes'.[56] However, we know that this did not happen for had they put on their masks, the area microphone would not have picked up their voices.[57] Why was this after the captain had said 'Yes'? Perhaps the 'yes' was merely an acknowledgement, as is commonplace in Japanese, that the captain had heard the request rather than an agreement to it. Perhaps this is why the flight engineer felt it necessary to make the suggestion twice and the co-pilot also made the request. Given the politics of a typical flight deck where the captain is very much the one who must be obeyed at all times, it is not inconceivable that the other two men were not prepared to put on their oxygen masks until the captain had done so. The significance that none of the three men put on their oxygen masks will be discussed elsewhere.

A further complication is that, during this discussion between the three men, JAL was attempting to contact JL123 through the company phone. The JAL half of the conversation was recorded on one of the channels on the CVR. The initial call came as the conversation about the oxygen masks began. A few seconds after the captain said 'Yes' for the second time, the company called again, and this time the co-pilot asked for the call to be taken. The flight engineer said he would take it, but then asks where something, which is not specified, is. What the captain says in response to this is unclear. It then appears as though there was some confusion as to whether the flight engineer was meant to be contacting one of the JAL offices in Tōkyō or Ōsaka, rather than taking the call from JAL. At 18:35:07 contact is finally made. JAL proceeds to say that they have been monitoring their emergency since it was declared at 18:26, when their location was 30 miles west of Ōshima. The flight engineer informs JAL that the R5 door is broken and that they are descending. JAL confirms this and asks if the captain plans to return to Tōkyō, using a mixture of English and Japanese. The flight engineer asks for this question to be repeated. JAL asks if they plan to return to Haneda to which the flight engineer replies, 'Wait a moment. We're now making an emergency decent' and 'We will contact you again in a little while… Again, we'll contact you again. Please keep monitoring us like this'. This is confirmed by JAL. In fact, at this time the phugoid motions were continuing, though less pronounced than before, and JL123 was not making a continued descent.[58]

Briefly during the communication with JAL and again at this point, the CVR picks up the communication of a purser. The exact contents are not clear, but it may have been in relation to how to use the oxygen masks.[59]

As mentioned above, the cockpit crew did not know much about the extent of the damage to the plane. They were aware of the loss of hydraulics, but with no monitors, sensors or equivalent to a car's rear-view mirror, they would not have known about the damage to the tail fin, other than what may have been sensed through trying to control the plane. If this was sensed, there was no discussion about it. Similarly there was no discussion about how to control the plane's direction through subtle adjustments to the engines on either side of the plane. For a period the cockpit crew appeared to be adapting to the demands and limitations of JL123 and the phugoid motion pattern became less extreme. During this time there was less communication than before, but what was said appeared to suggest a greater degree of team work taking place. For example, the captain spoke of 'putting the nose down', to which the co-pilot agreed. This would suggest the two were working together to control the pitch of the plane – which, without

Hiroyuki Ishii 石井裕之 *(59) · Sadaaki Ishikawa* 石川貞昭 *(52) · Rokurō Ishikura* 石倉六郎 *(41)*

hydraulics, must have been difficult. The captain even suggested to the co-pilot that he use both hands.[60]

At 18:38:31, the flight engineer asked if he should drop the landing gear. The co-pilot also repeated the question. Given that the plane was still at over 20,000 feet at this stage, the reason for this question is not clear. However, whatever the motivation, the captain said that he should not drop the landing gear.[g] Despite this, barely 25 seconds later, the flight engineer asks 'Shall I lower it slowly by alternating?' This time the captain says, 'Yes, wait a moment'. By this stage the plane was passing to the north of Mount Fuji and heading towards the city of Ōtsuki. The captain says something at 18:39:50 which is not clear, but appears to include the word 'lower'. At 18:40:22 the flight engineer confirms that the landing gear is down, which is acknowledged by the co-pilot. Around this time the phugoid motion ends and the plane enters what appears to be a more controlled descent down to around 7,000 feet. The conversation at this time is also relatively limited, with most of what is said being statements such as 'lower the nose', 'power' and 'heavy'.[61] The report suggests that 'heavy' was reference to 'the considerable steering force which would have been required… but it was impossible to clarify the reason'.[62] Given the lack of hydraulics and the amount of physical effort required of the captain, 'heavy' was probably a very appropriate term to be used. Some additional evidence relating to the amount of physical effort that was endured by the captain will be discussed elsewhere.

At 18:44:46 the flight engineer asks whether they should extend the flaps. The captain replies, 'It is too early'. The flight engineer asks whether it is too early, and the captain says that it is. The co-pilot then asks whether the landing gear is down, apparently forgetting that he had confirmed hearing the flight engineer say that it was down some four minutes earlier. The flight engineer confirms that it is down. At 18:45:37 JL123 was contacted by Yokota US Air Force Base to let them know they were ready for an emergency landing and informs them of the radio frequency to be used. The captain replies to say that 'Japan Air 123 uncontrollable', despite the fact that the plane had apparently been in its most stable phase since the explosion. ACC seems to be surprised to receive contact from the plane, having had no response to their previous contact five minutes earlier. ACC then asks whether they want to contact Haneda, but the captain requests that they maintain this radio frequency. Although ACC does not appear to hear the request the first time, it is confirmed after the captain repeats his request. At the same time Yokota Air Base confirms that they are ready and provides details of how to contact them. At this point the captain says, 'This may be hopeless'.[63] During this time the plane completes a 360 degree loop, with Ōtsuki just north of its centre, before heading east towards Haneda.[64]

At 18:47:10, the captain contacts ACC to ask for the radar vectors to Haneda and Kisazaru. ACC confirms they should maintain a heading of 090 for Runway 22 and then asks whether they can control the plane. The reply is that it is 'uncontrollable'. Around this time, the plane turns northwards rather than maintaining a 090 heading. At this point in the cabin area, a purser announced that the altitude was lower so that oxygen masks would soon no longer be needed. The next part of the recording is lost, due to a further communication from ACC, asking for JL123 to change its radio frequency, being recorded on the same channel. The continuation of the purser's message indicates that preparations were now being made for an emergency landing. This would mean that passengers were being asked to take up the brace position, with their head on the back of the seat in front. Those with babies were asked to hold them tightly. Passengers were also asked to check that their seat belt was done up and that the table was in the upright position. Meanwhile, having acknowledged the change in radio frequency, the captain called out, 'Mountain'. Instructions about which way to turn followed. Again the captain said, 'mountain' and further instructions were given. A few seconds later the captain repeats his concern, 'We're going to hit mountains' and says that maximum power should be applied.[65]

Uncontrollable

At 18:48:00 the plane was heading north-westerly towards the Japanese Alps, directly west of Ōme, at an altitude of just 6,800 feet and travelling 230 kt.[66] Further communication attempts from Yokota Air Base were ignored. Similarly, a further communication attempt by JAL was ignored. By now the attention of the cockpit crew was on trying to regain control of the plane, which was again in a phugoid motion. The tension in the cockpit was clearly rising, with calls to 'drop the nose', 'increase power', 'max power' and 'right turn' coming in quick succession. At 18:49:39, the captain said 'It's no good' as the plane began to stall and he asked for maximum power to be applied once more.[67]

At 18:50:27 the captain said '*Gambare*', which the official report translated as 'stick with it'. The comment sparked further activity from the co-pilot and flight engineer. Although apparently giving the plane maximum power, and saying that the controls were in their full position, the captain pointed out that they were 'losing speed'. Ten seconds later the captain appeared to add, 'You'll have to control pitch with power', although it is not clear that he says the word 'pitch'. However, given the response of the co-pilot and flight engineer it does seem probable that this is what he said. It is the first verbal indication of them understanding that power adjustments were a way to deal with the problems of phugoid motion. The speed is confirmed as being 220 kt.[68]

By 18:51:03 the plane had climbed to 9,600 feet, but its speed was only 190 kt having come out of the second of two rises and falls,[69] and the plane was descending again. The captain asked for the nose to be lifted. The co-pilot then asked what to do about the flaps. The flight engineer asked whether he should lower them, but the captain said they would not lower. The flight engineer said that they would go down by alternating and the captain acknowledged this. Now the captain gave instructions to lower the nose and seemingly asked the co-pilot to concentrate on that and to use both hands. Once again more power was asked for. The flight engineer then confirmed that the flaps were being extended. Again the captain asked for the nose to be lowered, adding that the co-pilot should 'push'. Thirty seconds later the flight engineer confirmed that the flaps were being extended. Again, twice, the captain asked for the nose to be lowered. Then, at 18:52:50 the co-pilot asked the captain if he should take over. The response is not clear, although the report suggest that he asks him to.[70]

At 18:53:03 the plane is at an altitude of around 13,100 ft and is doing 180 kt.[71] A few seconds later the captain asks for the nose to be raised and then for power to be increased. At this point ACC contacted JL123 again, to which the captain replied, 'Uncontrol. Japan Air 123. Uncontrol'. This was confirmed by ACC. ACC then asked JL123 to change the frequency of its signal, to the one which was requested six minutes earlier by ACC. Although both times JL123 had confirmed the request, it was not until this time that the flight engineer confirmed that the frequency had been selected.[72] The time was now 18:54 and the plane was at about 11,000 ft and doing 220 kt. At this point, JL123 was contacted by APC for Haneda Airport. The flight engineer asked for confirmation of their position. The response, in English, was 'Your position… 5… ah… 5, 45 miles north west of Haneda'. The flight engineer's response revealed the problem with the information: 'Northwest of Haneda… how many miles?' This time the information from APC was clearer, 'Yes, that is correct. On our radar you are 55 miles northwest of Haneda. Oh, 25 miles, ah, to the west of Kumagaya'. This information was confirmed by the flight engineer who passed it on to the captain and co-pilot. The plane was now at 11,300 ft and doing 180 kt. APC contacted JL123 again saying that it would contact them in Japanese and that Yokota was also on standby. APC then asked for JL123's intention. There was no response.[73]

Meanwhile in the cabin, according to Yumi Ochiai, most passengers had been putting on their life jackets and she, an off-duty JAL purser herself, and other pursers were helping passengers with this. Then passengers started to adopt the brace position. Some still had their oxygen masks on. 'From the front I heard a girl scream'.[74]

As Kawamura highlights,[75] there is still some uncertainty about the final path of JL123. But, based on the most likely path, any passengers on right-hand side of plane might have been able to make out the mountains running along the Gunma and Nagano prefectural border in the fading light. They may have seen Mount Osutaka in the distance and Mount Takamagahara closer to them. On this mountain they may even have seen an, as yet unnamed, densely wooded, ridge.

At 18:55:43, APC again contacted JL123, but in English. 'Japan Air 123. Japan Air 123. If reading, your radar position 50 miles, correction, 60 miles northwest of Haneda. Northwe… ah.. 5 mile, 50 nautical mile northwest of Haneda.' But as this communication was being sent, the scene in the cockpit of JL123 was becoming desperate. The plane had entered a steep dive – the steepest of its 32-minute ordeal and was dropping from just over 10,000 feet towards 5,000 feet.[76] Yumi Ochiai's statement continues,

> Then we started going down rapidly. It was a really rapid descent. It was upside down. It felt like my hair was upside down. It felt like the hair on both sides of my head was being pulled back. It probably wasn't really like that, but that's how it felt. It was frightening. I was frightened. Please don't make me remember any more. It was a fear you don't want to remember.[77]

Taeko Itō 伊藤妙子 (17) · Kazumi Itō 伊藤和美 (13) · Makoto Itsuki 一木允 (56)

Meanwhile in the cockpit, the co-pilot demanded, 'Flaps up, flaps up, flaps up, flaps up'.

Captain:	Flaps up.
Co-Pilot:	Yes.
18:55:56	
Captain:	Power, power, flaps.
Flight engineer:	We're going up.
18:56:04	
Captain:	Raise the nose. Raise the nose.
18:56:09	
Captain:	Power.

At this point the automatic ground proximity warning system (GPWS) began its alarm together with the words, 'Sink rate. Pull up. Pull up. Pull up. Pull up'. In those final few seconds one of the three men, thought to be the captain, said something, but it is unclear what it was.

The FDR shows that the plane did pull out of this steep dive and started to regain altitude again. As it did so, the right wing hit a single tree, known as the *ippon karamatsu* ('the single larch', *Larix kaempferi*),[h] and the outermost engine (Engine 4) came off.[78] The plane was also at a steep angle as it attempted to turn right to avoid the bulk of the mountain in front of it. The plane could not gain altitude fast enough and the right wing sliced through the top of the ridge, creating a 'U-shaped cutting'. The force of the impact shook the remaining engines from the wings as well as other parts of the tail from the plane.[79] The plane was now powerless. It continued lifelessly across the ravine and the pitch of the plane fell and the altitude dropped. The plane was now travelling at 264 kt and was almost upside down.[80] The plane broke up as it impacted against the ridge. Those at the front of the plane would have been killed instantly as they were subjected to a force of 'as much as hundreds of G'.[81] The rear of the plane continued its forward momentum, down the side of the steep ridge and came to a rest at the bottom of a gulley.

The time was 18:56:26.

Hiroyuki Iwai 岩井宏之 *(43)* · *Miyako Iwaki* 岩城美弥子 *(43)* · *Shinji Iwakiri* 岩切伸司 *(34)*

4 Response and responsibility

> They are so afraid the world will think the Japanese have not been able to handle the situation.[1]

The previous chapter set out the salient points relating to JL123's flight and ultimate demise. This chapter explores the way in which the crash was responded to. There are three key aspects which this chapter will look at in turn; first, there are the search and rescue (SAR) activities that took place in the immediate aftermath of it becoming clear that JL123 had crashed; second, the investigation into the cause of the crash; and third, issues relating to responsibility.

Trying to find JL123

Just before 19:00, JL123 vanished from the radar screens. With mountains potentially blocking signals, there may have been some who expected, or hoped, for JL123's image to reappear on the screen. Some may have even hoped that JL123 had managed to make some sort of emergency landing. However, the lack of communication meant that most would have been acutely aware that the plane had crashed and that the focus now needed to be on ensuring that SAR activities could begin. For SAR the first question to be dealt with is determining the location. As we have already seen in Chapter 3, there was a degree of vagueness about the exact location of the plane even when its image was visible and could be tracked on the radar.

The Air Self-Defense Force base in Hyakuri Air Base had also been tracking JL123's progress since its emergency transponder had been activated.[2] With the plane now off the radar, it would be logical that the SDF should launch one of its planes to pinpoint the crash site as quickly as possible. However, a plane has to be given permission to take-off. Living in a post-9/11 world it seems incredible to think that the steps to get this permission were not being undertaken prior to the eventual crash. Given that the captain had declared the plane as uncontrollable, surely a plane should have been scrambled prior to JL123 even crashing. A SDF plane should have been shadowing JL123 ready to shoot it down before it reached a point where it could endanger a large numbers of lives on the ground. This would also require political involvement in the decision making process.

Keiko Iwakiri 岩切圭子 (33) · Daiju Iwakiri 岩切大樹 (3) · Tadahiro Iwakuni 岩国忠弘 (43)

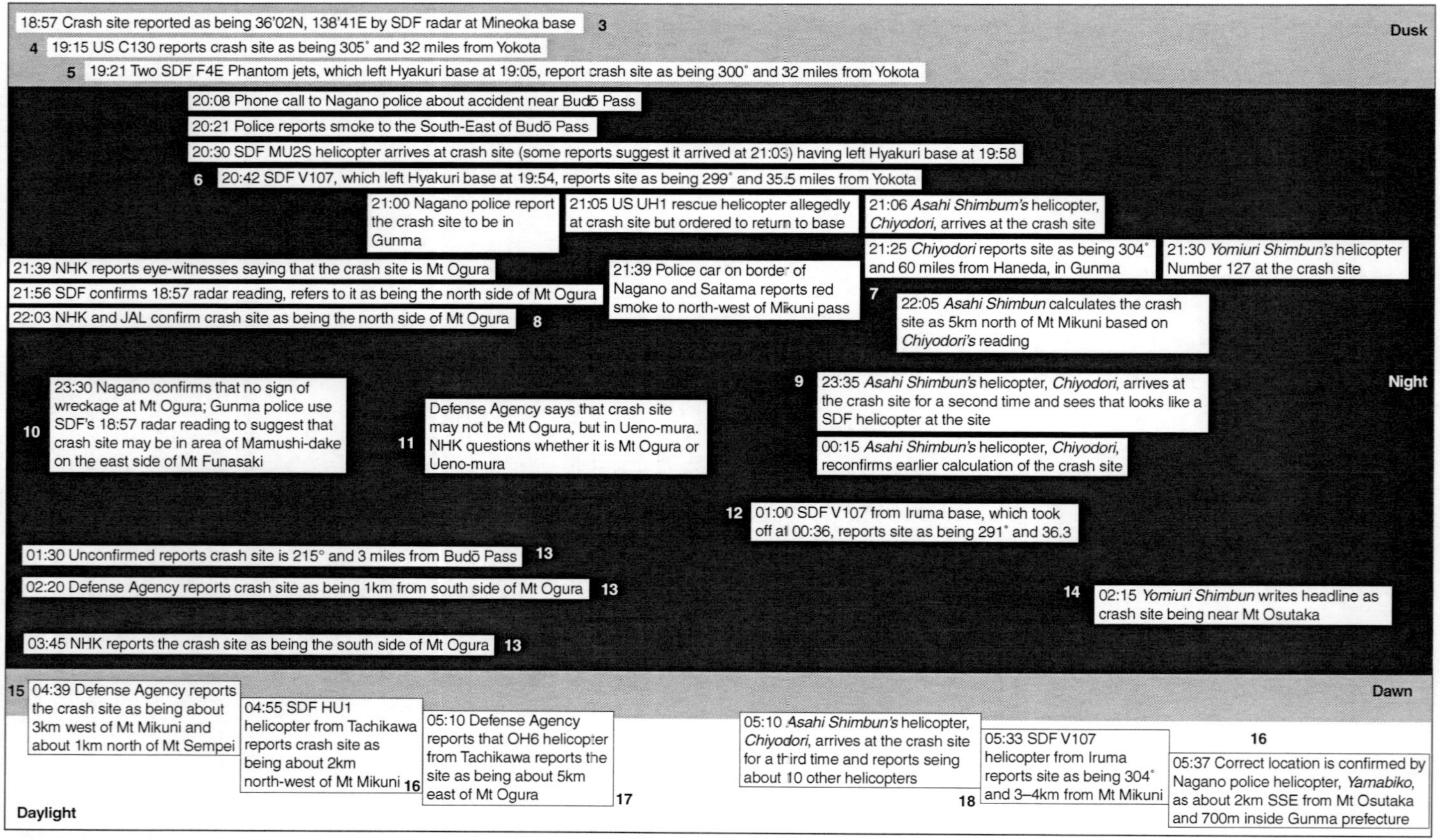

Figure 4.1 Finding the crash site – timeline

Figure by the author, based on Yoneda 2005:17, 55–57

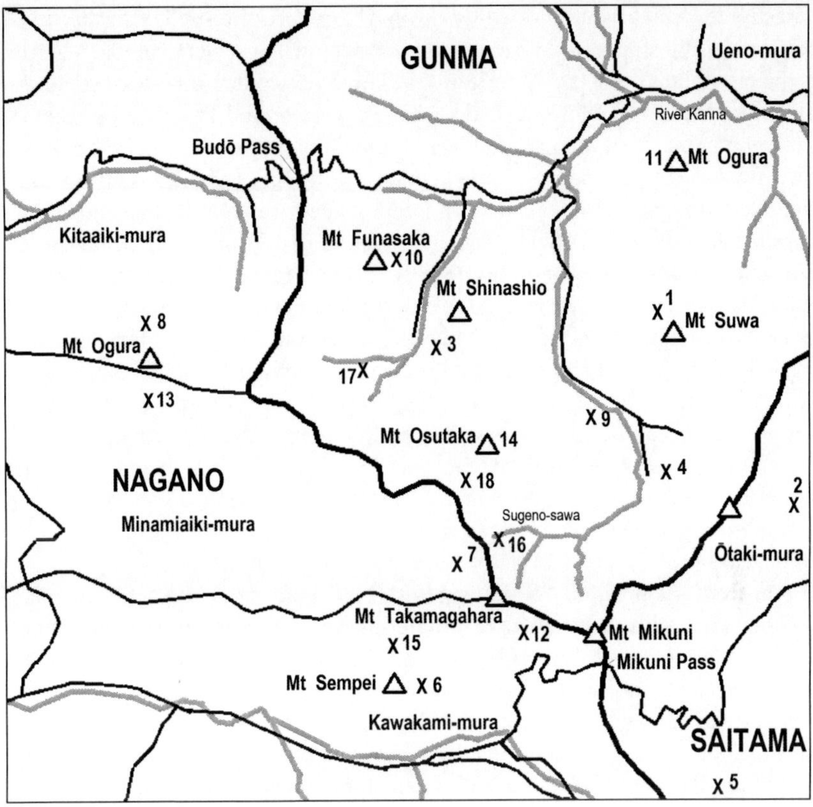

Figure 4.2 Finding the crash site – a map of the area

Map by the author. Based upon Yoneda 2005:76; Kawamura 2004:256 and information in Figure 4.1 in this book.

Note: In addition to the locations noted in Figure 4.1, the figure also shows the location of the crash estimated by ACC (point 1) and by Yokota Air Base (point 2) based on the readings when JL123 vanished from their radars.

As it was, two F4EJ Phantom were scrambled from Hyakuri at 19:01 and set off towards the probable crash area without waiting to get permission from an organization such as the Ministry of Transport.[3]

The SDF were not the only ones trying to find the location of the crash. The media were aware of the unfolding tragedy and also began their hunt for information. Most of the media's role in reporting the crash will be covered in Chapter 7, but it is worth noting here that by around 21:06 the *Asahi Shimbun* had managed to locate the crash site,[4] and the *Yomiuri Shimbun*'s helicopter arrived around 25 minutes later (see Figure 4.1). JL123's progress had also been monitored by an American Air Force C-130 Hercules returning from a training drill in Okinawa to Yokota Air Base.[5] After contact was lost with JL123, the C-130 diverted to the crash site, which is confirmed by the AAIC report.[6] In 1995, a further twist to this part of the story was added when a former American servicemen, Michael Antonucci, claimed that an American Air Force helicopter was directed to the crash site by the C-130, which he had been the navigator on.[7] The C-130 ultimately had to continue to its destination as it was getting low on fuel, but not before it had apparently successfully helped the helicopter reach the crash site. With the helicopter above the crash site, preparations were made for men to be winched down to the site and look for survivors. However, according to Antonucci, the helicopter was ordered to withdraw and return to base, as the Japanese rescue team was on its way.[8] In accordance with this directive, the helicopter withdrew and returned to base, whereupon those on board were ordered not to speak of the incident again.[9] The degree to which these particular events took place will be discussed in Chapter 8. What is pertinent here is that the US Air Force was apparently aware of the exact location of the crash site within the first hour of JL123 crashing.

In addition to searches from the air, land-based searches had begun and other information was coming in (see Figure 4.1). At 20:07 a policeman, responding to a call from someone living in Kawakami-mura, reported from his patrol car at Budō Pass (see Figure 4.2) that he could see white smoke in a south-easterly direction from his position.[10] Knowing with hindsight where the crash site was, his account fits. However, as Kawamura points out, it seems very unlikely that this 'white smoke' was connected with JL123, particularly as the policeman made no report of seeing any planes flying over 'the smoke'.[11] Due to the topography of the area and the darkness by this time, to be able to see smoke emanating from the crash site about 8.5 km from his position is improbable to say the least. A sighting from Mikuni Pass on the Gunma and Saitama prefectural borders, just 3 km from the crash site, is probably more reliable. That the report suggested that the smoke was red, reflecting the on-going fire at this time, would also support this.

Figure 4.1 summarizes the various suggested locations. As can be seen by dawn on 13 August there was still uncertainty about the location. At 04:39 a further update on the crash site was provided by the Defense Agency but even this turned out to be incorrect. Another SDF helicopter estimated the position at 04:55. Although this appears to have been the correct location, the Defense Agency made a further announcement at 05:10, giving a different, incorrect location. Finally, at 05:37 the correct location was confirmed by Nagano Police helicopter '*Yamabiko*'.[12, a] By now the sun had risen and the full extent of the crash could be seen. Media helicopters were also on the scene and broadcast the pictures across the nation. It was a horrific sight (see Figure 4.3). The V-shaped scar on the mountainside spoke of the size of the impact. The burnt trees were testament to the fire which had subsequently raged, which had luckily not spread and become a forest fire.[b] What was largely missing was the plane itself. Only one wing gave any hint that this was the site of a plane crash. On the wing itself were the letters 'JAL', although part of the 'J' was missing (see also Chapter 8). Even after the ANA flight NH58 mid-air collision in 1971 the front of the plane remained recognizable, much like the iconic shot of the Pam Am 747 in the Lockerbie disaster.[c] In fact, parts of the plane had survived the crash, but were hidden amongst the trees in Sugeno-Sawa.

Noriaki Kanbayashi 神林展明 *(33) · Hiroshi Kashima* 鹿島弘 *(64)*

Figure 4.3 The crash site at dawn, 13 August 1985

Photograph courtesy of *Asahi Shimbun*.

After the crash site was located, it still took about five hours for the SAR teams to reach the site, as will be discussed further below. First, let us just consider some of the salient points relating to the time it took locate the crash site. If one considers the time-line of main sightings and variety of location suggestions summarized in Figure 4.1, the obvious question to be asked is why did it take so long to accurately pinpoint the crash site? Given that the photograph produced by the *Asahi Shimbun* clearly shows that they found the correct location (see Figure 7.1, page 150), why did others struggle to do so? Indeed, the *Asahi Shimbun* was so perplexed by the continuing debate over the location during the night of 12 August that they sent the helicopter out a second time to reconfirm the location, getting there at 23:35, by which time the number of fires had reduced and another helicopter was also on scene.[13] Putting aside any 'conspiracy theories' about why the state may not have wanted the US forces on site or the correct location to be known, there appear to be two fundamental, yet significant, factors which may have contributed to the problems.

The first problem was maps. Maps are rarely accurate. Maps are flat, whereas the planet is not. Yet most of us rely upon flat images, whether it be a printed atlas, map, or computer screen. But the image of the world that such projections tend to produce are inaccurate. The most well known and well used of these is the Mercator projection. This map is so hideously inaccurate that I believe that it should be banned – particularly from educational institutions. The Gall–Peters projection, which attempts to accurately reflect area is an improvement, though contains its own problems. When dealing with a more local level, the curvature of the Earth becomes less significant, but the underlying principle – that the map is a representation drawn for the convenience of users – tends to remain. Accordingly, the level of detail can vary significantly depending on the anticipated needs of the user. An Ordnance Survey map, as used in the UK, has a high level of detail. Maps in Japan rarely have this. In relation to JL123, a significant point is that maps do not show the location of overhead power lines. The concern that helicopters may inadvertently hit one of these was given as one of the reasons for not initiating a significant SAR operation during the night of 12 August,[14] despite this seemingly not being a concern for the media helicopters. A British Ordnance Survey map would have revealed whether this would have been an issue. It seems ridiculous that the SDF is not aware where such potential obstacles are in the country which it is meant to be able to defend from foreign invaders. Just as ridiculous is that this reason for not using helicopters would appear to suggest that nobody thought of contacting a power company to check about the location of such power lines. In fact, even today there is little evidence of power lines near Osutaka-no-One.

While Figure 4.2 highlights the distance that Osutaka-no-One is from the centre of Ueno-mura, it cannot demonstrate the nature of the topography in the area. Ueno-mura is referred to as 'Japan's Tibet'[15] and 'Gunma's Tibet'[16] and it is easy to understand why. The village itself is hard enough to access, but the crash site is even harder. Even though a number of improvements have been made to the road over the years (see Chapter 6), even today it is not an easy drive. It is still 20 km from the main road in Ueno-mura and the drive can take around 40 minutes. The road follows the river and streams up from the village to the crash site. At times it is hard to see the sky as the mountainsides come close together and the trees and other vegetation are very dense. The mountains have very steep sides. If it were not for the vegetation growing on it, the mountainsides would be described as cliffs in places. When one remembers that the network of paths that exist today at Osutaka-no-One were not there in 1985, it is possible to appreciate just how difficult it would have been both finding and accessing the crash site. Furthermore, whilst most adults can cover 1km in around ten minutes, covering the same distance in most parts covered in Figure 4.2 could easily take several times larger due to the steepness of the slopes and the vegetation covering them. Each additional kilometre that the search team had to negotiate due to the inaccurate reporting of the location becomes highly significant.

Another factor that might have led to the mis-locating of the crash site could have been the scale of the maps and the units of measurements being used. Some measurements were provided by the SDF in nautical miles, whilst others gave longitudinal and latitudinal positions. Yet, people in Japan, tend to use metric measurements in everyday life. Consequently, each time a position was reported back, those trying to then pinpoint this on a map would have had to, at some point, convert the information to kilometres for passing the information to the search teams. There is an opportunity for an error, particularly given the conditions under which people were working.[d] That the distances were seemingly referred to as 'miles' rather than 'nautical miles' may have caused further confusion.[17] The problem of conversion could have then been further compounded by the scale of the maps being used. That some maps did not have longitude and latitude written one was a further problem,[18] whilst those that did were not consistent in their positioning.[19] The scale-related problems can be seen on the TV news broadcasts' maps which showed most of Japan, presumably so that viewers could relate it to their own home locations. The result was that mountains which are several kilometres apart appeared to be right next to each other.

Figure 4.2 also highlights an important point relating to locations and administration. The crash site is within Ueno-mura, but it is many kilometres from the centre of the village. There was no chance that anyone living in central Ueno-mura could have seen the plane or heard the crash. Yet, the crash site is less than 4 km from habitations in Kawakami-mura. So why was it that Kawakami-mura did not become the centre for operations? There are two apparent reasons for this. First, there was no direct road link between the village and the crash site. Although it is also true that there was a considerable hike from the final point on the road from Ueno-mura to the crash site in 1985, it was more straightforward than the trek from Kawakami-mura. However, perhaps a more significant point is that Kawakami-mura is in Nagano prefecture, whereas the crash site is in Gunma prefecture. Keeping all operations within Gunma, where the flying distance to the nearest city that could handle the next stage of the recovery process was also less than the equivalent in Nagano, was perhaps understandable. However, if it were not for this system, a shorter road for families could have been built from Kawakami-mura for remembrance visits in subsequent years, although this would have meant that other memorials would have needed to be located differently also (see Chapter 6).

Village, town, city and prefectural boundaries in Japan are clear, other than where a border dispute exists,[e] and the administration of any location is consequently known. Having grown up in the countryside in England, I find this odd. Every morning I woke to the sight of Caer Caradoc. The idea that it may be administered by the local village is incomprehensible to me. It is within Shropshire – a county, which is broadly equivalent to a prefecture in Japan. Road signs give some indication of which town or village I am in, but in rural areas even this can little significance, as can be seen by the number of those who live in locations 'near' a particular village or town. Taxes and other administrative functions occur at a county or borough/unitary authority level, as well as at the national level. It is the county boundary (which in the case of Shropshire includes part of the boundary between England and Wales) that is of any real significance. Given the discussions which have gone on in Japan in relation to municipality mergers, I cannot help think that many of the problems could have been overcome if the system was more alike that found in England – although this system is not without its own flaws. Regarding Osutaka-no-One, therefore, under an English set-up, it would be merely acknowledged that it is within Gunma prefecture. In 1985 Ueno-mura had a population of 2,083, yet its area is some 181 km^2.[20] This is about 40 km^2 bigger than Kawasaki which has a population of nearly 1.4 million, or about the same size as the unitary authority area of Warrington (population approximately 195,000) in England.

Yūta Kashiwabara 柏原勇太 (6) · Yukihiko Kashiwazaki 柏崎幸彦 (32)

The apparent desire to classify everything also extends to the naming of mountains. This is, of course, not unique to Japan. Throughout the world many peaks have names. Variations sometimes occur due to language differences, particularly when mountains straddle national boundaries, or due to the imposition of English (or other colonial) titles. In the case of JL123, there was an apparent need to pinpoint the crash site according to the name of the mountain. As can be seen from Figure 4.1, various names were suggested during the night. At no point was the name Mt Takamagahara mentioned, it seems. This name is not well known. Yet, Mt Osutaka also was not mentioned and, based on my own examination, it did not generally appear in atlases and maps available in 1985, suggesting that it would not be well known. Seemingly, the adoption of 'Osutaka' owes much to the power of the media, with this name being used both by the *Yomiuri Shimbun* in its headline of 13 August and also in some television reports that morning.[21] In fact, the ridge is on Mt Takamagahara. So, although the site is frequently just referred to as 'Osutaka', it is more correctly known as Osutaka-no-One (the ridge on Osutaka), despite being on another mountain. There is a further issue to be aware of which relates to the meaning of 'Osutaka-yama'. Ueno-mura has had many 'Osutaka-yama' over the years – that is 36 mountains in Ueno-mura have been referred to as 'Osutaka-yama' – mountains (*yama*) where hawks (*taka*) were caught during the Tokugawa Period (1603–1868) for the Shōgun.[22] With no definite or indefinite particles in the Japanese language, the difference between 'an Osutaka-yama' and 'the Osutaka-yama' is not clear. Although it is not certain whether Takamagahara was ever an Osutaka-yama, for many the name is not significant.[23] Given how inaccessible these mountains remain even today, I find it amazing to think that hawks were being caught on these mountains many centuries ago. In a similar vein, when looking down on mountains in Siberia during the flight between London and Tōkyō, I find the thought that some of these mountains may never have been walked on by humans as incredible as the thought that someone may have walked on some of those mountains.

Considering the timing of the crash, it is worth noting just how dark it can get. I find driving in the darkness in Japan, particularly in the countryside, much more oppressive than in the UK. Light pollution from urban conurbations and how much scattering material there is in the atmosphere is a well-known problem. It is also one which is much greater now than in 1985, and naturally much more significant in urban than rural areas. The difference in latitude between where I live in the UK and Ueno-mura, on the other hand, has no significant effect.[24] On the moonless night of 12 August 1985, the burning wreckage would have produced some light in the immediate vicinity and could have acted as a beacon for those approaching from the air in the initial hours after the crash. However, for those approaching on foot, where other mountains would have blocked any view of the fire, the only light would have been from their torches. It would have been difficult to gain an understanding of what was around them. I took light readings at Osutaka-no-One when I stayed overnight there in August 2010. By 19:30 the light reading was 0 Lux. By comparison, in the UK, when the sun had set at the same time and the moon was similarly not an influencing factor, I got readings no lower than 0.05 Lux in a city and 0.01 Lux in a rural location at any point during the night. Even when there is just a little light it is possible to make out the difference between the ground and sky – for example by making out the outline of a building or trees. On Osutaka-no-One I could not even see my own hand at the end of an outstretched arm when the reading was 0 Lux.

In thinking about the SAR, another factor which cannot be overlooked is that the sheer scale of the accident may have lead to mistakes being made. Indeed, the official report concluded that the time taken to locate the site was 'justifiable' under the conditions.[25]

Survivors are found

By around 09:00 the SAR teams were arriving at the main crash site.[26] Due, in part, to the problems locating the crash site, various teams approached the site from different directions. The SAR teams were not the only ones on site as the media were also present (see Chapter 7). There are suggestions that others had also reached the crash site by this time (see Chapter 8). In the UK, for example, the standard procedure would have been to contain the site and to ensure that there was only one entry and exit point site so that people coming and going, as well as any items being removed could be logged.[27] This did not happen for JL123. Of course, the way in which the site was found, let alone the nature of the site itself, was a complicating factor, but the fact that *izoku*, among others, were able to enter the site, despite being prohibited to do so (see Chapter 7), is somewhat concerning.

Once the SAR teams reached the crash site the focus was expected to be upon retrieving the remains of the victims, their belongings and the critical parts of the plane for investigating the cause of the crash. But, at 10:54, among the carnage and noise of the site, a survivor, Yumi Ochiai, was found.[28] The SAR teams then focused on releasing this person from under the wreckage as well as checking for other survivors. Over the next ten minutes a further three survivors – Hiroko Yoshizaki and her daughter, Mikiko, followed by Keiko Kawakami – were found.[29] All four were female and had been sitting near the rear of the plane (see Figure 3.5). Medical personnel were transported to the site by helicopter, arriving at 12:30,[30] 90 minutes after the survivors had been found. The survivors were then carried further up to the ridge to a clearing so they could be taken to safety. Although Keiko Kawakami was taken up by an SDF soldier (Figure 4.4), Hiroko Yoshizaki was taken up on a stretcher, which span round wildly whilst being lifted.[31] By now it was already 13:29 – two and a half hours had passed since they were found and some nineteen and a half hours had passed since the crash.[32] The helicopters transported the survivors to Fujioka, although initially going to Ueno-mura, so that they could be taken to hospitals.[33] The news of survivors caused much excitement. For example, I have seen a clip of the TV coverage of the high school baseball tournament taking place at Kōshien where the news was displayed on the scoreboard and was greeted with cheers and applause. For those *izoku* travelling to Gunma, the news that there were survivors brought hope,[34] but

Figure 4.4 A survivor is lifted from the crash site

Photograph courtesy of *Sankei Shimbun*.

for most it was false hope which further compounded the tragedy. The situation was not helped when there was an apparent double-counting of the survivors in the afternoon.[35]

The hope is that survivors can bring some insight into the cause of the accident itself. In the case of JL123, this hope appears to have been further exaggerated by the fact that one of the survivors, Yumi Ochiai, was an off-duty JAL purser. The assumption appeared to be that as an experienced flier, she would be able to provide more insightful comments than other passengers. Consequently once her condition was stabilized, attention turned to her account of what happened.

Before discussing the survivors' accounts in the next section, there are some points to be aware of. First, Yumi Ochiai's recovery was hampered by being together with two of the survivors, Hiroko and Mikiko Yoshizaki. As a JAL employee she was linked with the loss of life – including other members of the Yoshizaki family – which a JAL accident had caused. Ochiai's health, particularly her mental state, improved after the Yoshizakis were move to another hospital on 21 August and she was given her own private room the following day.[36]

That all the survivors were female and sat near the rear of the plane was a feature of the crash. Does this mean that sitting near the rear of the plane is safer? Are women better able to survive accidents? With two of the survivors being children, is it also the case that children are more likely to survive accidents? Statistically there is little evidence to support any of these ideas. Every crash is different. There are numerous examples of planes impacting upon the rear part of the plane first and so reducing the survivability of the crash for those at the rear. There are also numerous cases of men also surviving major aviation accidents. Whilst there have been some very notable cases of children surviving crashes,[f] there is no evidence to suggest they are more likely to survive an accident. There are many factors involved in whether someone survives a crash or not. For the four that *ultimately* survived JL123, they were lucky that these factors came together. The emphasis in the previous sentence is pertinent as the survivors stated that they heard, or even talked with, other survivors after the plane crashed.[37] The exact number is not known and there is little to allow one to estimate the number (see Chapter 5). One thing appears certain; had the SAR teams reached the scene earlier, there would have been more survivors.

Another point to note in relation to the survivors is that eyewitnesses are not always as helpful as one would assume. In the case of JL123, the plane was seen by two people working in a field as it flew over the village of Kawakami-mura shortly before it crashed. Their description of what they saw hints to some of the problems when dealing with eye witness accounts. 'At about 18:55 a plane which was larger than a Cessna flew over from the Saitama direction.'[38] Given the height at which JL123 would have been as it passed over them, the noise of the engines and its size would clearly have been recognizable as a large passenger plane. To make a comparison with a small Cessna is odd. But more puzzling was a report from the Nagano police of a sighting in Kitaaiki-mura where the plane was identified as the JAL logo was seen on the rear stabilizer.[39] Given that much of this part of the plane was missing, particularly the part with the logo,[40] this report seems inaccurate. Many eyewitnesses claim to see fire when a plane is coming down, but with notable exceptions such as the Air France flight AF4590 Concorde crash in 2000, planes generally are not on fire prior to crashing. The brain's recording of memories is not perfect and we have a tendency to mix up the order of events. There may also be a tendency to allow the events as we expect to have seen them, based on what has been seen in films, for example, to influence an account. Although small children do not have the vocabulary to provide a detailed account, it may be that older children, who might have been less exposed to media versions of the type incident they need to describe, could make good eyewitnesses. In the case of JL123, this means that the focus may have been on the wrong survivor.

Hirotsugu Kawaguchi 河口博次 (52) · Hisashi Kawaguchi 川口寿 (40)

Probable cause

Let us now focus on the 'probable cause'. This term is the one used at JAL's Safety Promotion Center and is a translation of *suitei genin*. The official report's concluding section is '4.2 Cause' in which it states:

> It is estimated that this accident was caused by the deterioration of flying quality and loss of primary flight control functions due to rupture of the aft pressure bulkhead of the aircraft, and the subsequent ruptures of a part of the fuselage tail, vertical fin and hydraulic flight control systems.[41]

The word 'estimated' (*suitei*) is used heavily throughout the 'Conclusions' part of the report, but the nuance is that calculations have been made and a conclusion reached, which is perhaps different from the nuance of 'probable cause' as a term.

Naturally, there was a desire to quickly find the cause of the accident, although the initial focus seemed to be on retrieving the bodies from the crash site, potentially at the expense of preserving key evidence relating to the cause of the crash. At the time there were over 600 Boeing 747s in operation around the world.[42] Only a few weeks earlier another 747 had been lost, so a second accident involving the plane, where pilot error did not appear to be a factor, was a major concern.[g]

The initial focus was on the R5 door as this had been identified as a problem during the communications between JL123 and the ground. Some suggested that any problem related to this door may be connected to the accident in 1978, but a JAL spokesman pointed out that the accident in 1978 was unconnected to the R5 area and that the damaged area had been completely repaired so it was 'Impossible to think that this could be connected to the this crash'.[43] Although the English version of the official report does not make this clear (see below), the door was soon located at the crash site and was found to be, other than damage probably sustained during the impact, in good condition.[44]

On 13 August, at 18:55 – almost exactly 24 hours after the crash – a piece of the plane's vertical stabilizer was found in Sagami Bay. That some of the tail had come off was also confirmed by a photograph taken of the plane in its last few minutes. The photograph was taken by amateur photographer, Keiichi Yamazaki, as it went over his house. Realizing that it was lower than a plane would normally be, that it was moving strangely from side to side and making an odd noise, he got his camera and took a picture.[45] As anyone who has ever taken a picture of a plane flying overhead will know, what appears like a large object in the sky can end up being a depressingly small in the final photograph. Figure 4.5 shows the original picture and an enlargement of the plane. What is clear from the picture, at least for an expert (see Chapter 8), is the extent of the damage to the tail. After this picture was analysed, it was calculated that between 55 per cent and 59 per cent was missing.[46] Although these facts pointed to probable contributing factors leading to the crash, they could not explain what triggered the damage and the sequence of events. Perhaps the accounts of the survivors, who were sitting in the rear part of the plane near where the damage was, would provide some clues about the cause.

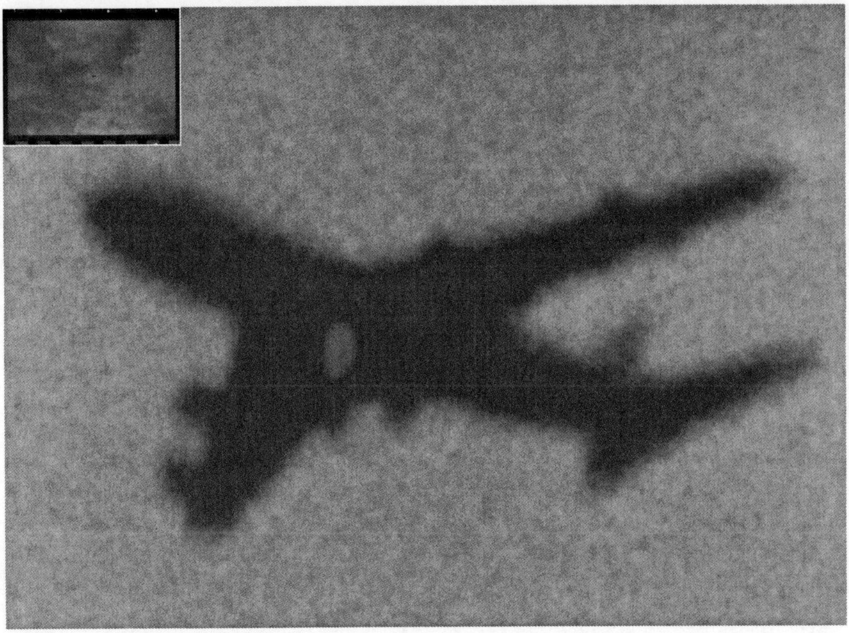

Figure 4.5 The last picture of JL123

Pictures courtesy of Keiichi Yamazaki. Inset shows the original photograph; the main image shows only the aircraft.

Indeed, the first clue to the 'probable cause' appeared to come from the 'Ochiai statement', which appeared in newspapers on 16 August. This statement referred to a variety of factors which are common in cases of depressurization. However, the statement also included reference to the plane experiencing 'Dutch roll' and Ms Ochiai being able to see that vent holes were open. When Ms Ochiai was interviewed by Yoshioka (see Chapter 3), she confirmed that at the time of the statement she was not aware of the term 'Dutch roll' and she was not able to see any vent holes from her seat.[47] For these reasons, it is best to put this 'statement' to one side.

On 16 August some of the media was reporting suggestions that the rear bulkhead of the 747 had failed and that this had led to rapid depressurization of the cabin. The air that escaped from the cabin had critically damaged the plane. The reason for this failure was linked to the accident JA8119 had had in 1978. According to some sources, it was the initial investigation by the Americans from Boeing and FAA at the crash site that led to a suggestion that the rear bulkhead was the problem (see Figure 4.6). Apparently, the visual inspection of the rear bulkhead led them to realize that there were some rivets missing, although there was a feeling that perhaps the Americans had access to information that the Japanese did not have.[48] A diagram of this was then put on the rear of some *meishi* by Ron Schleede and given to Hiroshi Fujiwara of the official accident investigation team.[49] As Shun Takeda of the official investigation team said in a subsequent

Figure 4.6 The rear bulkhead at the crash site

Picture courtesy of Mainichi Shimbun.

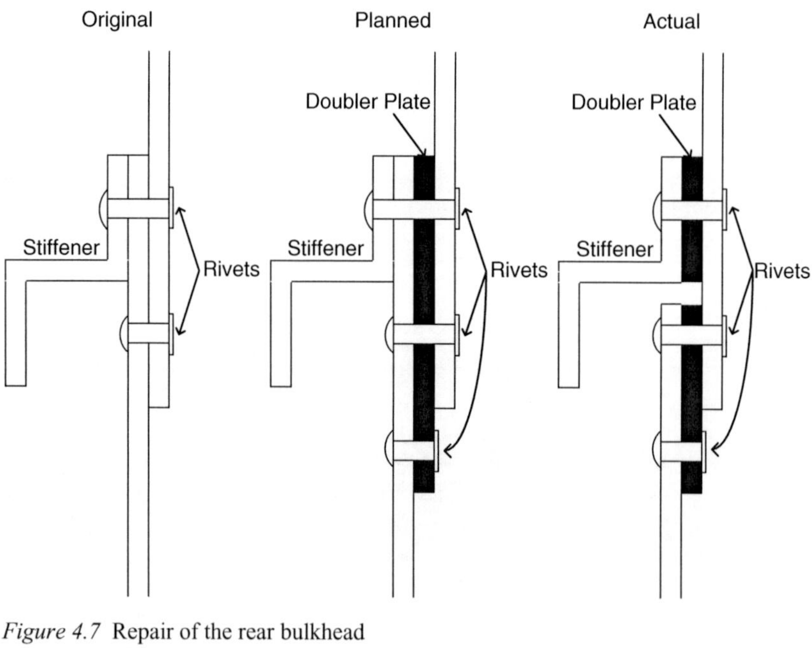

Figure 4.7 Repair of the rear bulkhead

Figure by the author.

interview, once Boeing announced their belief that the faulty repair was to blame (see below), the 'process [of making the report] became easier'.[50] However, there appear to be some inconsistencies between the accounts and with fact. Whilst one account suggests that the American team was only able to access the site two weeks after the crash,[51] newspaper accounts suggest that the team first accessed the site on 16 August. But either way, it is still inconsistent with the fact that, as will be discussed in Chapter 7, the idea that the rear bulkhead was the key to the cause was the headline in one newspaper on the morning of 16 August.

Figure 4.7 shows the areas of concern relating to the rear bulkhead. It is estimated that a faulty repair reduced the overall strength of the bulkhead and that this would reduce the number of cycles it could withstand to around 13,000.[52] Flight JL123 was JA8119's 12,320th cycle since the repair.[53] According to the theory, the bulkhead gave way and the air in the cabin escaped into the tail section. Although there is a pressure relief door designed to cope with such a situation, it is suggested that its size was not able to cope with the amount of air coming out from the cabin area. Consequently, pressurized air travelled up the hollow tail. The tail reached a point where it could no longer withstand the pressure differential or force of air, and broke off. In breaking off, the APU and the hydraulic lines were severed, leading to the plane's loss of hydraulics. It was the combination of a lack of hydraulics and a lack of vertical stabilizer which sealed JL123's fate.

Problems with the Official Report

Takeda's words cited earlier point to a central criticism of the final report – that it appears to have been written in order to maintain a fixed conclusion rather than to investigate all major possibilities and to assess them. As a piece of research it is terrible. If one of my students submitted a dissertation of this sort, they would fail. It does not respond to the survivors' accounts. It does not deal with the concerns relating to whether there was rapid depressurization. The survey to find remains of the plane in Sagami Bay appears to be insufficient. The report is also poorly written. Let us consider each of the points in turn.

Although one would not necessarily expect survivors' testimonies to appear in a final report, parts which are relevant may be included in the USA, for example.[54] Given that Ochiai was cabin crew, Weir would have expected more of her testimony to be used had she been a US survivor of a US crash.[55] That it was not is perhaps in part due to the inconsistencies between the 'Ochiai statement' and the much more detailed interview she did with Yoshioka. However, it is also worth noting that Ochiai was also interviewed by George Seidlein of the NTSB on 27 August 1985. In this interview, which was made public after the file in which it was contained was passed to Hideo Fujita, she stated that her ears did hurt at the time of the explosion and while white mist appeared, that there was no sense of the mist or air moving.[56] All of her evidence is consistent with there having been depressurization. But there is no reference to any sensation of air being sucked out of her. There was no rapid movement of air. This all raises question marks about whether there was *rapid* depressurization, which is a prerequisite if the official report's conclusion is to remain valid and this is perhaps why her comments were not included.

I do not think anyone disagrees that there was depressurization of the cabin, but depressurization is not localized; it would have affected the whole plane. Standard procedure in such a case is for the pilot to rapidly reduce the plane's altitude to a level where it would be safe for people to breathe unaided by oxygen masks. This process can be terrifying for passengers, already alarmed by any explosion associated with the start of the depressurization, as there might not be an announcement, automated or from the cockpit crew, as to why the plane is descending at such a steep angle (as noted by the media in the case of FR9336 below). There are examples of such incidents. In relation to this occurring on a Boeing 747, the two that are particularly relevant are China Airlines Flight CI611 in 2002 and Qantas Flight QF30 on 25 July 2008. In the case of CI611, the plane is thought to have suffered explosive decompression on a part of the structure due to a faulty repair after a previous incident.[57] However, unlike JL123, CI611 lost structural integrity and broke up in mid-air. Although the aircraft loss happened quicker than that of JL123, it is worth noting that the force of the decompression was strong enough to rip clothes off a large number of the victims, a fact that has been seen with other similar incidents. In the case of QF30, there were no fatalities following the rapid decompression and the plane made a safe landing. Following the explosion at 29,000 feet, the pilot put the plane into an emergency decent, reaching 10,000 feet some five and a half minutes later. As a result of the rapid decompression a number of passengers suffered nausea (related to problems of not being able to get their oxygen mask to function properly and the resulting oxygen deprivation at that altitude). Many also said that their ears popped, and that they saw mist and debris in the cabin.[58]

However, it should be noted that experiences can vary. Although rarely mentioned in English aviation accident books, for the Japanese, particularly those with an interest in JL123, a point of comparison came barely 14 months after JL123. The incident occurred in Japanese airspace, although the carrier was Thai Airways International flight TG620 which had left Bangkok and was bound for Itami via Manila. As it flew over Tosa Bay a bomb was detonated in the rear toilets of the Airbus 300-600, which had 233 passengers and 14 crew on board. There was rapid decompression after a hole was blown in the rear bulkhead. The plane made an emergency descent and landed safely at Itami. There were no fatalities, but five passengers and three crew were badly injured and needed to be taken to hospital, and 101 others had minor injuries. After the explosion, passengers without seat belts on had been lifted out of their seats, food and trays flew around the cabin, those walking in the aisles banged their heads on the ceiling.[59] The difference between the experiences of these passengers and what the four survivors of JL123 described after the explosion on that plane were apparent. However, although on a smaller plane, 16 of the 168 passengers on board Ryanair flight FR9336 on 26 August 2008 were hospitalized due to ear ache following a decompression of the cabin.[60] While others could have suffered less extreme ear ache, this case does suggest that there are variations in how individuals experience and react to decompression.

Those who disagree with the official conclusion point to a number of problems. First, had it been rapid depressurization then the pilots would have known to take the plane down to 13,000 ft straight away.[h] The pilots did not do this. Indeed, after the explosion, JL123 stayed above 20,000 ft for about 18 minutes and above 13,000 ft for a further three minutes (see Figure 4.8). Above 20,000 ft anyone without oxygen would have suffered from acute hypoxia, leading to unconsciousness within five to twelve minutes.[61] Yet JL123 was above this altitude for between two and three times this length of time. The only people on board who did not use oxygen were the cockpit crew (see Chapter 3). The official report suggested that the lack of talking between the cockpit crew was an indication of them beginning to suffer from hypoxia, also noting that it was estimated that they could have flown for about 18 minutes at an altitude above 20,000 without a mask.[62] Clearly there are differences in views about how long someone can survive without a mask at this altitude. However, we also know that another reason why the captain did not say much was that he was grinding his teeth.[63]

Another problem pointed out by opponents of the rapid depressurization idea is that the survivors do not mention there being any significant wind in the plane or movement of the white mist which appeared at the time of depressurization. That the report includes mention of the latter point whilst ignoring the lack of the other is another area for concern relating to the official report in its apparent desire to only present information which suited its needs. Had there been rapid depressurization, it is claimed that the air would have travelled at a typhoon-like speed of 60m/s,[64] and air would have been sucked from passengers' lungs, causing great discomfort. Survivors would surely remember this and comment on it. Those who wrote *isho* may have commented on it as they commented on the mist, but they did not (see Chapter 5). Similarly, as the plane had a hole, perhaps the temperature should have stabilized to the ambient temperature. At that altitude the temperature would have been close to –40 °C. None of the survivors mentioned feeling cold, despite only wearing T-shirts and shorts. However, with the engines still on, warm air would have still been blowing into the cabin which may have accounted for the lack of feeling of cold.[65]

As already discussed, some of the tail of the plane was found floating in Sagami Bay, but much of the tail was never recovered. It is assumed that most of it came off the plane at the time of the explosion. If this was the case then it should have fallen into Sagami Bay. An underwater investigation was conducted in November 1985[66] and no further parts of the plane were found. However, the search covered only a limited an area and so many believe that evidence, which may be critical in helping to disprove (or even prove) the existing 'probable cause' theory, could still be at the bottom of Sagami Bay.[67]

There are also issues with the way in which the report was written. The conclusion is relatively comprehensible. However, the same could not be said for the main contents. It is almost as though they were either written by different people or with different audiences in mind. The way in which the main contents are

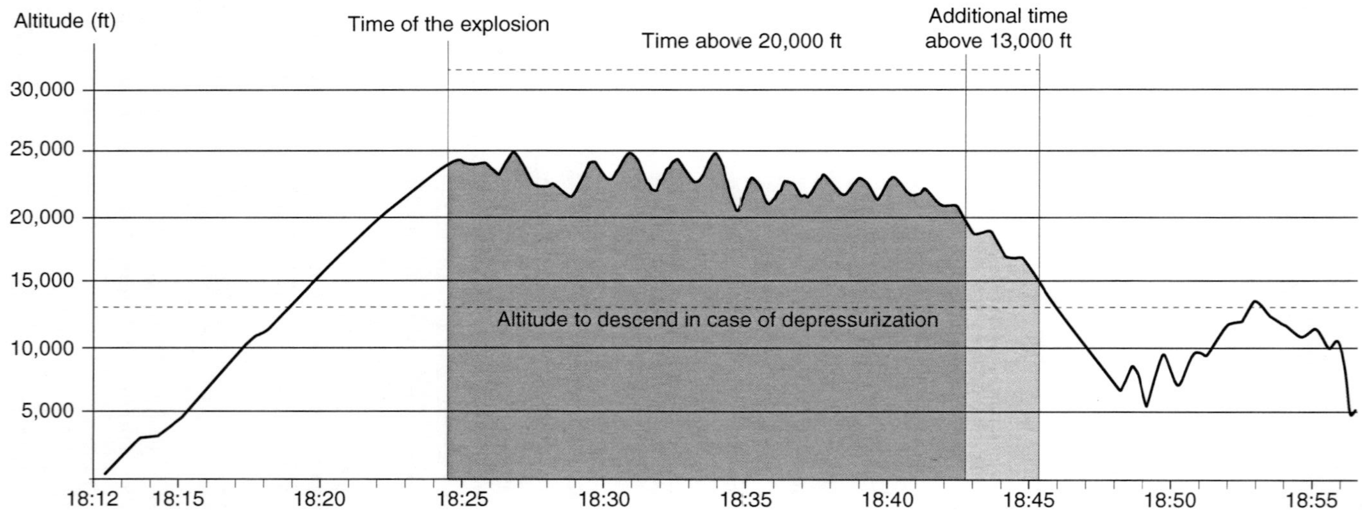

Figure 4.8 Changes in altitude of JL123

Figure by the author based on AAIC 1987a:Appendix 5.

written, many *izoku*, for example, have complained that only experts in avionics could understand them.[68] There are also some minor inconsistencies between the Japanese and English versions of the text.[i] Further, as Yoneda and Fujita point out, the transcription of the CVR appears to have been done to fit with how the investigation team wanted to interpret the words, which at times were hard to hear. After the CVR was passed to Fujita and he then made it public, analysis of it revealed a number of instances where the original report's suggestions do not appear to fit.[69] Some *izoku* have also pointed out that, having concluded that the 1978 misrepair was at fault, the report does not go into sufficient detail about how this misrepair could have happened, why it was not discovered or what steps can be taken to ensure there is no repeat of it.[70]

So, if it was not rapid depressurization, what could have caused this accident? Putting aside the 'conspiracy theories' (see Chapter 8), the main suggestion put forward is 'flutter'. 'Flutter' is the process by which a vibration builds up and can lead to damage developing.[71] It is suggested that the tail suffered such damage and that when it subsequently broke, the process caused some minor damage to the bulkhead also allowing some air to be lost. This suggestion is not based purely on a rejection of the rapid depressurization theory. Analysis of the CVR reveals that it has picked up sounds – inaudible to the human ear – just prior to the explosion which may be indicative of 'flutter'.[72]

Chapter 3 has already pointed out that the CVR on JA8119 was longer than the usual 30 minutes and consequently it includes the sound of the explosion. Although this seems a piece of good fortune, perhaps the recording missed something critical that was going on in the minutes prior to the explosion. So the fact that the CVR begins only moments before the explosion, and that this explosion has been tied to the rupturing of the rear bulkhead, may have distorted the reality of what really happened on the flight. Had the CVR on JA8119 been 30 minutes long and so not recorded the sound of the explosion and the conversation that was taking place at that time, the investigators might have had to look at other evidence more carefully. Similarly, had the tape been longer, it would have included information on the missing period from the last communication with JAL and the time of the explosion.

There are many things which will remain a mystery about what happened on JL123. For example, after the purser got permission to allow a passenger to go to the toilet (see Chapter 3), did she get to the passenger before the explosion? Or was she able to converse with the passenger from her own seat? Were either of them standing at the time of the explosion? If they were, and if there was rapid depressurization, then they would probably have been knocked off their feet and injured. There were no reports to the cockpit of any injuries. Was this because there were not any or because they were not serious enough to warrant disturbing the cockpit in an emergency situation?

In 2008 I travelled on a Boeing 747 between Japan and Hong Kong and sat in the equivalent position to where Mr Ogawa was sitting (see Chapters 3 and 8). While the decision to take the picture of the inside of the plane (Figure 3.3) is one that has puzzled Mr Ogawa's son,[73] and the intent and contents of some of the other pictures he took is also the subject of some speculation (see Chapter 8), the question that struck me is why was there no picture of the damage to the plane? His seat would have been close to the damage. The noise emanating from the explosion and subsequent damage would almost certainly have caused many in the area to turn and look, just as Yumi Ochiai did.[74] Yet, not only are none of the passengers in Mr Ogawa's picture of the inside of JL123 facing backwards, he did not take a picture of the damage himself. His son says that he was given all of the photograph negatives back, so there is no indication that any pictures were deliberately withheld. Was it simply that from his angle it was not possible to get a good, meaningful picture of the damage?

We will return to other theories about what may have happened in other chapters.

Responsibility

Whatever the actual cause of the crash, the fact is that on 6 September, an article, based on information from Boeing, appeared in the *New York Times* saying that the faulty repair of JA8119 led to the crash. Boeing did not notify JAL in advance that it was going to do this, much to the displeasure of many JAL employees.[75]

Why would Boeing do this? If it really was felt that the faulty repair of the rear bulkhead was the cause, then there was little reason for Boeing not to make the statement. Making the statement would serve to reassure airline companies and passengers that JL123 was an isolated case and that there would be no need to ground the rest of the 747 fleet. Did this fact mean that Boeing was prepared to fall on its own sword? With so much at stake and with JAL being Boeing's largest customer at the time,[76] was Boeing was using the convenience of the faulty repair as a means to control the situation?

Although Boeing's actions also helped JAL, as it relieved the company of most, if not all, of the blame, I do not believe that we should necessarily see Boeing's actions as part of some conspiracy – either by Boeing acting alone or with other parties. There really could have been a belief that they had found the cause. Any special orders from Boeing following the crash to look issues relating to the tail of its 747 planes may indicate that the company was at least concerned that the cause was something other than the rupturing of the rear bulkhead due to the faulty repair. Discovering whether this happened or not is beyond the scope of this study, however. Regardless of Boeing's motives, it still should have remained imperative for the official investigation to check whether this was the probable cause or whether there could have been any other cause. This does not appear to have happened satisfactorily.

One of the issues with the official investigation is that there were effectively two investigations going on. One of these was that being conducted by the Aircraft Accident Investigation Commission (AAIC). In 1985 the AAIC was a part of the Ministry of Transport. The team that investigated JL123 was made up of an investigator in charge and 15 investigators. Thirteen technical advisers were appointed for particular research and six personnel from the Ministry of

Transport also helped. On top of this, other organizations also aided in 'the fact finding investigation'.[77] Among the organizations that aided the AAIC was a team from the National Transportation Safety Board (NTSB) in the USA. Although only brief mention of this is given in the report, the foreign input could have been key in finding the probable cause, as discussed above. There were also issues with the foreign help, for example, the team originally not being granted access to the crash site.[78] The AAIC published a draft of its report on 28 March 1986 and a hearing was held on 25 April. Issues relating to this hearing are discussed in Chapter 9. What the AAIC report does not clarify is that there was a simultaneous investigation being conducted by Gunma Prefectural Police. Their purpose was to ascertain whether a crime had been committed and, if so, who was responsible. As Yanagida points out, Japan tends to 'focus on finger-pointing'.[79] Gunma Police's work took primacy over that of the AAIC, and so the AAIC had to get permission from the police to borrow parts of the plane wreckage for their investigation.[80] Although this probably had no significant impact on the final outcome of either investigation, the concerns about how the process was conducted is one that remains 25 years after the crash. While the investigations were on going, the International Federation of Air Line Pilot Associations were so concerned about the leaking of confusing and contradictory information to the national press that they had to make a public statement reaffirming that the priority should be on finding the cause and not finding who was to blame. They also pleaded that representatives of Boeing and the NTSB should be allowed to participate more in the investigation.[81] Although NTSB representatives were taken to the crash site, some have described this as merely a 'courtesy gesture' and after that their input was not sought.[82] This would appear to contradict subsequent interviews which suggested that their involvement led to the discovery of the 'probable cause'.

Whatever the failings of the investigation system in Japan, and regardless of whether the faulty repair in 1978 was directly connected to the loss of JL123, there is one thing which cannot be escaped from – the Boeing 747 was not properly designed. The 747 was designed with a number of fail-safes in place. It is probably for this reason that it did have such an excellent safety record. Although the AAIC concluded that the fail-safes were in accordance with the standards required by the Federal Aviation Administration (FAA) of the USA,[83] it appears there were at least three fail-safes missing. The first of these was a means to prevent all hydraulics being lost in the case of a leak at the rear of the plane. This design error was subsequently corrected with the insertion of a valve as a consequence of what happened to JL123. The second design fault was not having a door across the access to the tail. Had this been in place on JL123, then air could not have rushed up the tail leading to its destruction. Again, this was corrected after JL123. The third fail-safe that was missing relates to the rear bulkhead itself. 'Certification requirements… state that the aircraft's structure must be able to maintain its integrity in an explosive decompression.'[84] Weir also points out that the 747 in Britain is 'illegal' as Section 2.1.3 of the British Civil Aviation Regulations state that 'the primary structure shall be designed to withstand any pressure differences which might exist between compartments and, in particular, to withstand the effects of sudden release of pressure'.[85] While the corrections made appear to support the 'tombstone imperative' (see Chapter 2), the fact that the faults were there at all suggests that there was a problem in the design process itself. There should have been designers playing a devil's advocate role, asking questions such as 'what would happen if this went wrong?' What happened to JL123, in terms of the rupturing of the rear bulkhead, was not that extraordinary. There had been many other incidents like this before. Had the 747 been designed properly, the JL123 crash would have been avoidable.

Naturally, many were critical of Boeing as a result of the faulty repair and its supposed consequences. For example, Shintarō Ishihara, who served as Transport Minister between November 1987 and December 1988 and who went on to become Tōkyō Governor, wrote in his notorious book *The Japan That Can Say 'No'*, 'Five hundred and twenty people died because Boeing workers were so incompetent or careless that they could not securely fasten a three-ply bulkhead. Such shoddy performance by a Japanese corporation is unthinkable'.[86] Of course, Ishihara was to be proven on many occasions subsequently that his assumptions about Japanese corporations were incorrect,[j] but the anger he felt was not limited to over zealous ideas about Japanese workmanship compared to American workmanship. The problem was, as Ishihara notes,

> under US law, an aircraft maker is not prosecuted for criminal negligence. The logic behind this is that rather than punish a few individuals, the public interest is better served if they candidly describe what happened and a recurrence is prevented. The US legal system assumes that unless people are given immunity, they will not tell the truth. Given the scale of the accident, that premise was unfathomable, not only to the families and friends of the victims but to most Japanese.[87]

Indeed, one of the lawyers who represented some *izoku* believes that had those who did the repairs to JA8119 been Japanese, 'there is no question that they would have been prosecuted',[88] but as it was there was no possibility of charges being brought, and Gunma Police did not prosecute. For many *izoku*, there were unanswered questions about the cause and feelings that those who were responsible for the faulty repair 'got away with it'. Although Boeing would end up paying the bulk of the compensation (see Chapter 5), no action was taken against the individuals who made the faulty repair. The whole affair was reduced to a financial transaction. Although there was seemingly no possibility of any charges being brought against Boeing or JAL, it did not stop many *izoku* from trying. In total, by 1987, 697 people representing 195 families made complaints, with over 3,000 non-*izoku* also filing charges with Tōkyō District Court.[89] One group of *izoku*, made up of around 70 Japanese families and ten non-Japanese families, attempted to use the American courts, but without success.[90] Thanks to the Freedom of Information Act in the USA, some found it easier to get information about the crash there than in Japan. This fact was not missed by the media and led to groups, such as the Japan Civil Liberties Union, pushing for a similar act in Japan.[91] Such an act was promulgated in 1999 and came into force in 2001. Many documents were destroyed by the government prior to the act becoming law.[92] Indeed, it was concern that some vital documents relating to JL123 were to be destroyed, that led the CVR to get into the public domain after it was seemingly passed from a ministry official to Hideo Fujita.[93]

The reaction of many of the *izoku* is understandable. Had I lost a loved one, I would want someone to vent my anger at. As their actions lead to death, what should the penalty be? As it was not murder, the death penalty would be too extreme. But would an imprisonment be sufficient? Would it even be appropriate? In the short term it may satisfy much of the *izoku*'s anger, but if we consider the desire of many *izoku* for there to be greater air safety (see Chapter 6), could such a sentence, and even the threat of such action be counterproductive? I understand the logic of why, in the case of an aircraft accident, it may be desirable for those with key knowledge to come forward without fear of repercussions so that similar accidents in the future can be prevented. This was not murder. It was a mistake that lead to death. If a worker makes a mistake, should you fire them or allow them to learn from the mistake, retraining them if appropriate? Does the scale of the mistake matter – particularly if it leads to a fatality – or does the principle remain the same?

In terms of taking responsibility, there is an additional incident that needs to be mentioned. On 21 September 1985 a JAL employee committed suicide. The exact reasons for the suicide are not known. He worked in the maintenance centre at Haneda, but had also been assigned to look after one of the *izoku* (see Chapter 5) and was also assigned to a department to look after compensation negotiations. His suicide noted 'I apologize by killing myself',[94] but for what he was apologizing for was not clear. Thanks largely to the work of Ruth Benedict,[95] who has been widely criticized by Japanese scholars for including her own value judgements in her work,[96] it is often suggested that Japan is a 'shame-based culture' rather than a 'guilt-based culture', but there were many in JAL who were concerned that the suicide would be seen as an indication of responsibility and guilt about the cause of the crash.[97] As Reader *et al.* note, Benedict 'seems to postulate that guilt and shame are entirely unrelated to each other, which is obviously contrary to the facts'.[98] Part of the issue is that 'guilt' can be viewed in different ways. For many Westerners it may be 'an inner problem for the individual', whereas for the Japanese, guilt 'shows itself most sharply when the individual suspects that his action will result in betraying the group to which he belongs... it is a function of human relations'.[99]

JAL President Takagi resigned on 14 August to 'take full responsibility'. But rather than stepping aside straight away, he remained in place until the time was appropriate for a successor to take over. From a Japanese perspective there was nothing surprising about this, and most of the media around the world also understood, but one American radio station contacted JAL to find out why Takagi was 'chickening out'.[100] Sixteen days after the crash, Takagi went to the Diet where he bowed deeply and offered his apologies again.[101] Haruta and Hallahan point out that 'when an organization makes a mistake that disrupts society, the large power

Mariko Kishi 貴志真理子 *(35)* · *Eriko Kishi* 貴志絵里子 *(7)* · *Satoru Kishi* 貴志論 *(4)*

distance in Japanese culture' demands that the top person takes responsibility and apologizes; something that they find less likely to happen in the US, for example.[102] It was not until December 1985 that Susume Yamaji, a close associate of the Prime Minister, took over as JAL President.[103]

Conclusion

This chapter has covered many areas relating to the response to the crash – from the initial SAR activities to finding the cause to discussing responsibility for the crash. In terms of using the responses to the JL123 as a means to understand aspects of Japanese and non-Japanese societies, discussions about the cause are to some degree peripheral. What is of greater concern is the way in which the actors reacted to the disappearance of JL123 from the radar screens and the way in which the subsequent SAR and accident investigations were conducted.

The opening words of this chapter were written by Lyman Cotten, then American naval attaché in Tōkyō, in relation to the Japanese response to the 1923 Great Kantō Earthquake. Their implication has seemingly continued to echo throughout time. What really happened in relation to the allegation that US help after JL123 was turned away will be further discussed in Chapter 8, but this chapter has already highlighted that, in many respects, the Japanese were not able to handle the situation. Why could this be? Is there something inherent in the Japanese that means they are unable to respond to a disaster? This clearly is not the case. There are numerous examples, as will be discussed further in Chapter 5, of those who responded extremely well in the wake of JL123.

Rather than people, the problem appears to be the system. I have read and heard many accounts about how individuals from JAL, the Gunma Police, and the SDF responded very well after JL123, but when looking at these groups as a whole, the criticisms are much more numerous. At an individual level, although having to comply with directives above, there are possibilities for people to reveal their own personalities. This can be a key component in dealing with *izoku*, for example. However when looking at the system, the apparent lack of someone who can take control and make bold decisions may have led to problems (see also Chapter 5). It is perhaps ironic that a country so noted for its group-based culture, that this very feature may be the greatest weakness in a disaster situation. Indeed, it may be the component that turns a tragedy into a disaster.

At this stage it is easy to look at what happened with hindsight and use logic and analyse actions. But in the confusion at the time, the likelihood for people to act irrationally and for mistakes to be made is great. It was an exceptional time. Having said this, some of those involved are paid and trained to deal with exceptional circumstances. So why was the SAR seemingly so poor? There seem to be a number of factors, none of which are necessarily mutually exclusive. First, there were concerns about the possible dangers caused by an isotope being carried on JL123 and it was not until 4:50 on 13 August that it was confirmed to pose no risk.[104] Given the apparent lack of plane visible, even at night when a fire was raging, perhaps the idea that there would be no survivors took the urgency away from mounting an effective SAR. But this still does not explain why the crash site itself could not be accurately pinpointed and then accessed. Perhaps it was a combination of the mistakes and problems discussed above due to issues relating to the maps, topography and the dark. Some suggest that the SDF were on site earlier than the recorded time and were removing evidence that would reveal that JL123 had been shot down (see Chapter 8). There were also clearly systemic problems, leading to what has been described as 'institutional paralysis'.[105] That the SDF had also not mounted an effective rescue after one of its own submarines sank a boat in 1988 would be indication that what happened in the wake of the JL123 was not an isolated problem.[106]

Although an emergency rescue team existed, it was never called upon for JL123. Unlike this rescue team, Gunma Prefectural police had no helicopter at their disposal, despite the mountainous nature of the prefecture.[107] When a military helicopter located the site, it took no action as the orders had gone to another unit.[108] Consequently, 'confusion about bureaucratic jurisdiction greatly delayed rescue efforts, and most probably cost a number of lives'.[109] Who should have taken responsibility for this? In the aftermath the divisions between the various groups involved were all too apparent.[110] What was lacking was central leadership. This is particularly ironic given that the Prime Minister at the time was Yasuhiro Nakasone, noted for his desire for decisive top-down leadership. Indeed, Nakasone had been on holiday in Karuizawa, in neighbouring Nagano at the time of the crash. But his involvement in dealing with JL123 appears pitiful. Looking through the diary of the Prime Minister's activities, as published in the newspapers, one can see that he did not visit the crash site or go to see the *izoku.* In fact, Nakasone did not go to Ueno-mura or the crash site until 4 November.[111] This is in marked contrast to Margaret Thatcher who interrupted her holiday in Australia to go to the scene of the British Airtours fire at Manchester Airport only ten days after JL123. The photographs of this featured in all the national newspapers in Japan. Although there may be times where such an intervention can be disruptive to the recovery process, a Prime Minister should surely be seen to be trying to do something positive. Nakasone sent flowers to the temporary morgues in Fujioka (see Chapter 5). He accepted the resignation of JAL President Takagi, whom he 'severely admonished for the allowing the crash to occur', adding that 'there was a lack of discipline among JAL employees'.[112] As Nakasone confirmed who the new President would be, he called the top executives of JAL 'irresponsible, arrogant men',[113] adding that it was time for the belief that 'the government will support us' (*oyakata Hinomaru)* needed to end.[114] Surely Nakasone should have been doing more than this. Van Wolferen argues more did not happen, as 'to do so would have created the unwanted impression that the government had taken symbolic responsibility for the event'.[115] Surely that is what a government is supposed to do.

Part III
The human cost

5 Counting the cost

Mariko, Tsuyoshi, Chiyoko,
Be good to each other and work hard.
Help your mother.
It's sad, but I'm sure I won't make it.
I don't know the cause.
It's been five minutes now.
I don't want to take any more planes.
Please *kami-sama* help me.[a]
To think that our dinner last night was the last time.
There was some sort of explosion in the cabin
There was smoke and we started to descend
Where are we going, what will happen?
Tsuyoshi, I'm counting on you.
Darling,[b] it's too bad that this has happened
Goodbye
Please take good care of the children
It's 6:30 now.
The plane is turning around and descending rapidly.
I am grateful for the truly happy life I have enjoyed until now.

Hirotsugu Kawaguchi *isho* (see Figure 5.1)

Let us return to August 1985. After the recovery of the four survivors, attention turned to retrieving the remains of the victims. The remoteness of the crash site meant that various decisions had to be taken about the way in which the bodies would be transported off the mountain and where they would be taken. Whilst the centre of Ueno-mura is not the closest municipality to the crash site, the crash site was within Ueno-mura. It was also most accessible from central Ueno-mura and so it was natural that Ueno-mura would be the base of operations. However, Ueno-mura is a very small village in terms of population and facilities. With over 500 corpses to be brought down from the mountain, there was concern about a lack of space for these bodies prior to their identification. Further, as the route from the centre of the village to the crash site had no proper road for much of its length,

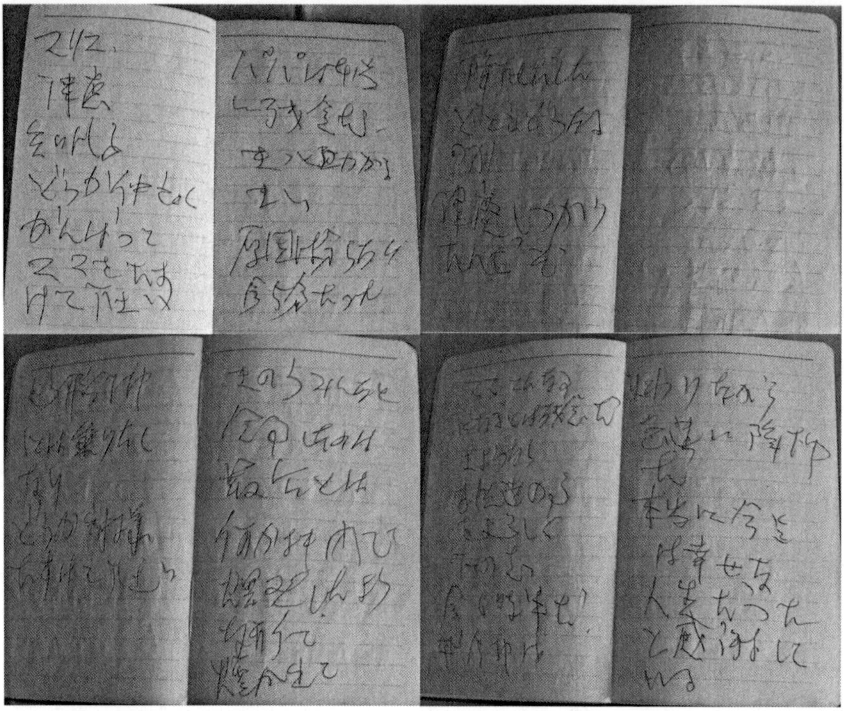

Figure 5.1 The Kawaguchi *isho*

Photograph by the author with permission from Mariko Kawaguchi.

and what existed was narrow and only passable by one vehicle, traffic going both ways was unimaginable. Transport to and from the crash site would largely have to be done by helicopter. With the road from Ueno-mura to the nearest city in Gunma also being twisty and time consuming, it was decided that it would best if the remains were flown directly to Fujioka.[1] There, they could be taken to one of the municipal buildings which would become a temporary morgue. As it was the school holidays, school gymnasia could be used as waiting areas for the families.

Once this decision had been taken, the SDF began constructing a heliport at the crash site, as the use of winches was slow and not without its own risks. The front of the ridge where the plane had impacted was clear of trees, wreckage and bodies, so was selected as the site for the heliport. Construction continued though the moonless night, using lights so that it would be ready the next day.[2] Consequently, no remains were transported off the mountain until the 14 August, when work on identifying the victims could begin. A second heliport was constructed between 15 and 16 August for larger helicopters.[3] Remains were wrapped in blankets and taken to the helicopters by SDF soldiers, who covered their mouths to protect them from the dust being blown around by the helicopters and to reduce the impact of the smell of the rotting corpses (see Figure 5.2).

Figure 5.2 The SDF at the crash site

Picture courtesy of *Mainichi Shimbun*.

If we consider the numbers of SDF soldiers who were mobilized to Osutaka-no-One, it is possible to begin to appreciate the sheer scale of the event. The number of SDF troops at Osutaka-no-One between 13 August and 13 October was about 52,000 (see Table 5.1). Including the number of other personnel that were also at the site, on some days in August there were several hundred people there.[4] It is important to stress that many of these soldiers did not leave the site at the end of the day, and needed to sleep there, amid the devastation, on slopes and in the open.[5] Rotations were set up so that teams stayed at the site for three to four days.[6]

Whilst the focus was initially on recovering the human remains from the crash site, it was also necessary to retrieve the close to 150 tons of plane for the investigations.[7] Kawamura claims that this process, other than for the FDR, CVR and rear bulkhead,[8] did not start until 17 September.[9] However, if one looks at the

Table 5.1 Number of personnel responding to JL123

		August	*September*	*October*	*November*	*December*	*Total*
Employed Fire Service	Dispatched to site	1,064	248	0	0	0	1,312
	On standby at site	1,340	80	0	0	0	1,420
	Employed fire service	2,404	328	0	0	0	2,732
Additional fire service	Dispatched to site	1,919	50	0	0	0	1,969
	On standby at site	2,979	0	0	0	0	2,979
	On standby home	29,885	0	0	0	0	29,885
	Additional fire service	34,783	50	0	0	0	34,833
Police	Search team	38,123	17,475	7,624	2,457	1,952	67,631
SDF	Search team	47,877	3,188	910	0	0	51,975
Prefectural Staff	Dispatched to Ueno-mura	344	55	2	0	0	401
	Dispatched to Fujioka	2,767	220	24	0	0	3,011
	Information collection	1,075	453	208	150	102	1,988
	Prefectural staff total	4,186	728	234	150	102	5,400
Municipality Staff	Information collection	5,589	742	21	0	0	6,352
Red Cross	Body identification	966	83	0	0	0	1,049
Doctors	Body identification	1,907	310	100	52	82	2,451
Total		135,835	22,904	8,889	2,659	2,136	172,423

Table by the author.
Source: Gunma-ken Sōmubu 1986:46

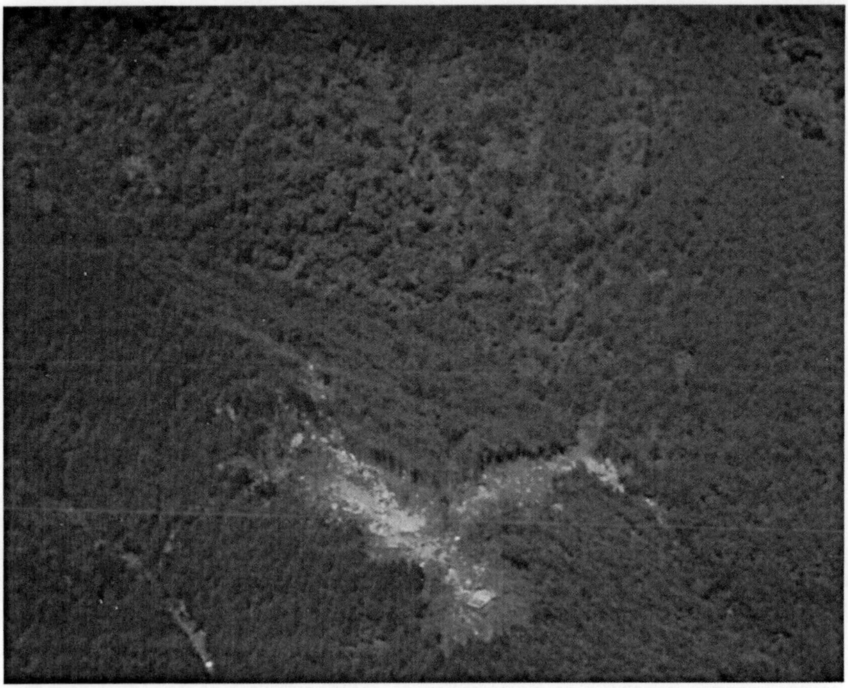

Figure 5.3 Osutaka-no-One from the air, 22 August 1985

Picture courtesy of Peter Mathews.

pictures taken of the crash site taken by *izoku* from helicopters above the site on 22 August (Figure 5.3), it is noticeable that the left wing of the plane, the only significant piece of wreckage visible from the air on the morning of the 13 August, is not there. Removal of the plane's remains was time consuming and expensive, costing approximately about ¥1 m per ton.[10]

Looking after the *izoku*

After the news of the crash broke, many *izoku* travelled to Itami or Haneda to get more information. There were also some demands that they be taken to the crash site. At this stage there was still hope that their loved one(s) would have survived, and so they wanted to be able to be reunited with them, whether in a hospital or another venue, as quickly as possible.[11] Many families carried with them changes of clothes for their relative(s).[12]

During the evening of the 12 August, buses, provided by JAL, left Ōsaka and Tōkyō, bound for the Japanese Alps.[13] Their initial destination was Kōmi town in Nagano.[14] Without the mobile communication to which we are accustomed today, stops at service stations were needed to provide opportunities to receive new instructions from headquarters. Other than those who had radios turned on for news updates, the buses were quiet, punctuated by the occasional sound of crying,[15] although, as is common in Japan, most tried not to show their emotions.[16] The buses from Ōsaka continued through the night before reaching their revised destination, Fujioka, at about 14:00, some 12 hours after they had left.[17] JAL also arranged a special flight on the morning of 13 August from Itami to Haneda for other families wishing to travel to Gunma.[18] The flight was also largely quiet, although some crying, from passengers and stewardesses, was to be heard.[19] Around 20 buses, with a police escort, then took the families onto Fujioka, where they arrived at about 16:00.[20] One of the buses had no air-conditioning or opening windows so the conditions were very unpleasant.[21]

Clearly, aspects of the support provided could have been better executed, but that should not overly detract from JAL's efforts. Although there may be an inherent belief that costs may be recuperated from insurance payments, the steps signalled that JAL was not going to stop to worry about financial concerns over what was considered appropriate action for the families.

By the time families were arriving in Fujioka, the scale of the tragedy was clear and that *izoku*, i.e. the relation of someone who had died, was the appropriate term to be used. A JAL staff member was allocated to look after each *izoku*. This role in Japanese is referred to as *sewayaku* (mediator), and in English has recently become referred to as being part of a 'care team'. With one exception, all of the *sewayaku* were male and aged 40–50, as 'Japanese people believe older males are better prepared than females or younger males to handle a stressful situation like the aftermath of an airline crash'.[22] A critical point to understand in relation to the *izoku* is what is meant by the term itself: family relations. A wife, a child, a parent – these are all *izoku*. A girlfriend, a boyfriend, a lover – these are not *izoku*. This is significant when it comes to getting information, support and who receives compensation (see below). Indeed, one woman whose fiancé died on JL123 turned to Hideo Fujita to get support and help in understanding the crash as she found that she could not use other channels.[23] The gay lover of another victim similarly had problems getting support.[24]

There is much literature about what different people involved in some way with JL123 did, but there are gaps (see Chapter 8). One group whose story has not always been told in such detail is that of the JAL staff. There were some reflections in the in-company magazine, *Ōzora*, but little mention is given in other sources, other than the fact-based novel *Shizumanu Taiyō* (see Chapter 8), which has limitations to using it as a source. I have had the opportunity to hear much from one JAL *sewayaku* (see Chapter 1). Whilst his situation was different in that he was non-Japanese looking after a non-Japanese family, around the time of his retirement, following a company interview for an in-house training DVD relating to JL123, he was told that much of what he said was the same as the Japanese. The main difference was that he said more. Many of the Japanese just did want not say what they experienced or 'dared not' talk about it.[25]

The main role for these JAL staff was to provide support for the *izoku*. They got information, found places to stay, helped with transportation and provided essentials such as food and clothing. With so many families travelling to Fujioka, often with several people from each family, accommodation and food were a major problem. There was simply not enough accommodation in Fujioka itself, so hotels had to be found in around 80 locations in Maebashi, Takasaki and elsewhere around the prefecture. Some were put up in *ryokan* in the *onsen* resort of Ikaho, 40 km from Fujioka, for example.[26] At the other extreme, I have come across a case of one *izoku* having to stay in a love hotel[c] due to the shortage of accommodation in the first few days.[27] It required 93 charter buses to shuttle *izoku* to and from their accommodation, the school gyms where they waited for updates, and other locations in Fujioka to identify their loved ones or their belongings.[28] As for food, as well as using local suppliers, JAL even transported in food from Narita Airport to help cope with the numbers.[29] But there were limitations to what JAL could do. JAL, which was being treated as a suspect by Gunma Police,[30] often found it hard to get key information and so was often not significantly better informed than the *izoku* themselves.[31] A further piece of support that JAL provided was that on 22 August, with many bodies not having been identified, JAL arranged for some *izoku* to be taken in helicopters and flown above the crash site.[32] Many then dropped flowers down on to the site. As can be seen from Figure 5.3, the site was cleared of SDF and others at the time and blue sheets were put over sensitive areas.[d]

The conditions in the school gyms were not pleasant. In Maebashi the temperature hit highs in excess of 30 °C on all days expect one through August.[33] With humidity being high, the effect of being in any shade would have been minimal. Japanese schools are relatively Spartan by many European countries' or North American standards, so there was neither air-conditioning nor fans until JAL provided fans after a few days.[34] Families sat around reading newspapers, watching TVs covering the latest news about the crash and recovery efforts, and looking at maps of the area (see Figure 5.4). Many families were also interviewed by the media (see Chapter 7). Charts at the front of the gym provided information about the location of the crash site and a seating plan showing which victims had been identified. Although there are references to some *izoku* interacting with each other,[35] they were generally focused on their own situation and desire to identify their loved one(s).[36] The 'communitas', which will be discussed in Chapter 6, was seemingly yet to develop although the seeds may have been sown.

Supporting the *izoku* was difficult. Chapter 4 has already discussed the suicide of one JAL employee. Several others suffered in different ways. Many *sewayaku* had to deal with abuse from *izoku*, and there are also stories of JAL staff being thrown out of restaurants in the days after the crash as the public sentiment turned against JAL.[37] Some of those in Fujioka ended up sleeping in the schools.[38] These employees had not received special training prior to this work; they were attempting to play a very difficult balancing act. They were representatives of JAL and JAL itself was seen as responsible for the crash – regardless of the actual cause. JAL employees saw themselves as being responsible for the situation the *izoku* were now in, but as they accompanied *izoku* into the morgues, they were also exposed to the sights that the *izoku* were. Although the JAL *sewayaku* had no personal connection with the victims, most of them will have been affected by

Figure 5.4 Waiting for news in a school gym

Pictures courtesy of Peter Mathews.

the sight of a dead body – particularly considering the condition of many of the victims. About ten *sewayaku* were told to stop on doctor's orders,[39] whilst another died.[40] *Sewayaku* did not receive counselling afterwards. Indeed, the lack of counselling generally is a major deficiency in the whole story of what happened in the wake of JL123 as the *izoku*, too, did not receive counselling.

It was not just JAL that was providing support to the *izoku* in the gyms. There were also volunteers from Fujioka. The main help that these, predominantly female, volunteers provided was to supply tea and *oshibori* (see Figure 5.5) and someone to talk to. *Oshibori* are dampened cloths used for wiping the face.[e] In the end, there were a total of 3,700 volunteers, with 388 being the highest number on any one day.[41] The volunteers came from 30 groups, such as the local women's association (*Shirengōfujinkai*), the PTA and the Chamber of Commerce and Industry (*Shōkōkaigijo*).[42] The acts of the volunteers were very gratefully received, and, as will be discussed in Chapter 6, the volunteers and *izoku* have developed a strong bond which continues to this day.

It is important to acknowledge the existence of these volunteers. There was a perception that the Japanese did not volunteer, reflecting the difference between a 'do to others as you would wish to be done to yourself' attitude and 'do not do to others what you would not want to be done to yourself' attitude. Such views probably exaggerated the reality, perhaps in an attempt to suggest a difference between 'us' and 'them'. Indeed, there is evidence to suggest that voluntary

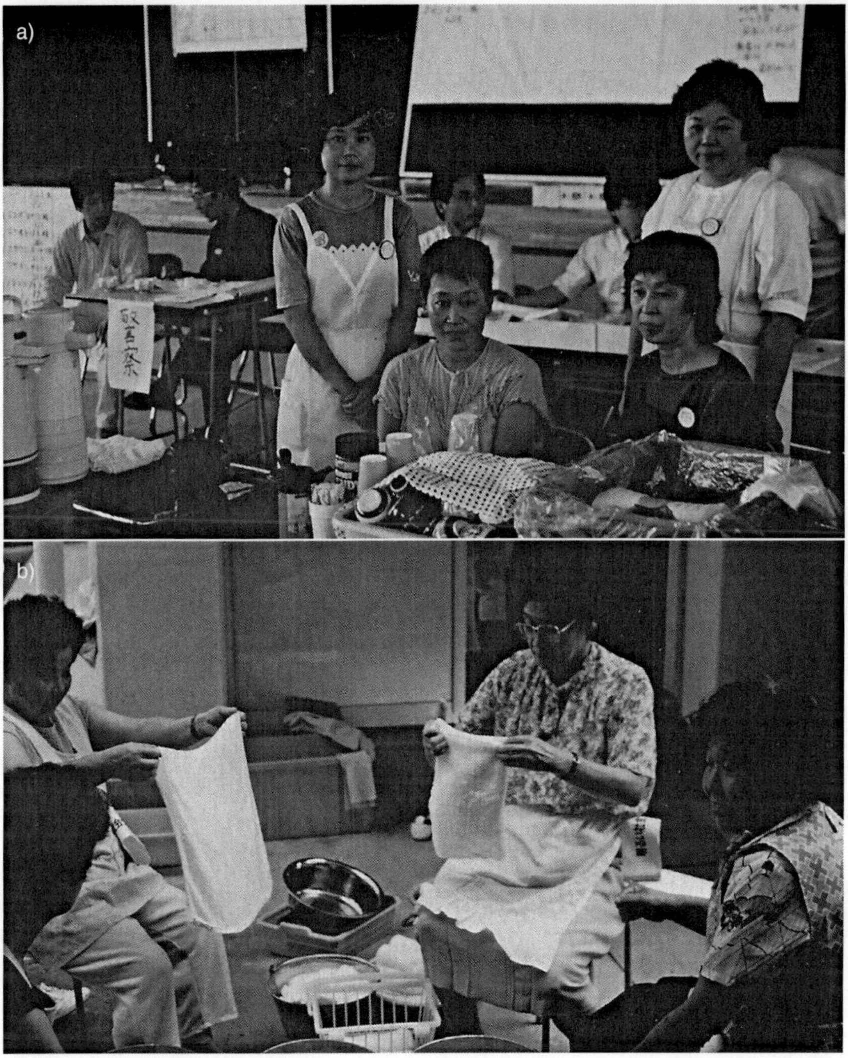

Figure 5.5 Volunteers in Fujioka

Pictures courtesy of Peter Mathews.

activities which gained popularity in Victorian England were a form of *tatemae*, or tokenism, rather than a real desire to do good.[43] There may even be elements of this in Japan. For example, those who gave hospitality to pilgrims were thought to get good fortune, whilst those who did not would be punished.[44] Whatever the differences between the motives and origins of volunteering in countries such as the UK and Japan, some, as Nakano points out, suggest that 'people in Japan possess only a superficial understanding of volunteering or lack the will to volunteer'.[45] It was because of this and the sheer numbers of volunteers that went to Kōbe and

Kazuko Kubo 久保和子 *(42)* · *Yumi Kudō* 工藤由美 *(24)* · *Masahiko Kuninaga* 国永昌彦 *(52)*

surrounding cities following the Great Hanshin Earthquake that many referred to 1995 as being the '*gannen*' (first year)[f] for voluntary activities in Japan.[46] Yet, while the activities that took place after the Great Hanshin Earthquake were important in many respects, volunteering had emerged in the 1970s,[47] if not earlier. Indeed, civil society was active in Japan in the 1950s, as can be seen with the response to the deaths of members of the *Daigo Fukuryū Maru* fishing boat, which was caught up in the Bikini Atoll atomic test on 1 March 1954. A movement led by housewives in Tōkyō collected 50 million signatures.[48] Another movement, again led by housewives, ensured Tamagawa in western Tōkyō was conserved.[49] One must not underestimate the importance of such movements. As Scott points out

> civil society is the primary locomotive for creating social capital, defined as features of social organization that promote trust, norms and networks that can improve the efficiency of society… Voluntary associations that bring people together for various reasons are the most important building blocks of social capital. This social capital will have political inputs.[50]

The way in which the JL123 *izoku* became a part of this process will be looked at in Chapter 6.

Other than the volunteers in Fujioka helping the *izoku*, it is worth mentioning some other groups of volunteers involved with JL123. First, there are the SAR teams from Ueno-mura. This included firemen, who supplement and support the main prefectural fire service, and members of the *Ryōyūkai* (hunting club), who traditionally hunters were but are also involved with looking after the land. These groups were instrumental in helping the SAR teams find the crash site. I find it a bit disturbing that their local knowledge rather than the expertise of the military in SAR activities was not only seen as being so critical, but also naturally accepted as being the case in the accounts in many books and documentaries. There were others who apparently not only went out in an attempt to locate the crash site and try to help those there, but may have actually got to the site before the SAR teams (see Chapter 8). Such actions, given the problems of traversing the topography in the area, particularly at that time of day, let alone the knowledge that a crash site was likely to be a traumatic and potentially dangerous place, indicate that the idea that the Japanese do not volunteer has always been overly simplistic. Indeed, other than the volunteers covered in this section, this chapter also points how doctors and dentists gave up their time to help in the wake of the crash.

In addition to the support from volunteers and JAL, the public also made a variety of donations. Seventy types of *imonhin* (comfort articles) were donated.[51] These included 9,648 instant *rāmen*, 26,294 cans of juice, 2,810 towels, 2,000 sweet breads, and 50, 360 packet boxes of *umeboshi* (dried plum).[52] In addition to the donations by individuals, companies also made direct donations. For example, one supermarket company donated ¥1 m worth of cup noodles and juice.[53] There were also people who came from neighbouring Nagano prefecture with 100 chrysanthemums (see also Chapter 6).[54] There were perhaps only 65 donations of money, totalling ¥5,357,000,[55] which would suggest that this form of donation was not popular.

Identifying the victims

Many of the bodies were found in and around what had been the rear of the plane, which went down the ridge to Sugeno-sawa. Across the rest of the site there were few complete bodies, with bags needing to be used for placing body parts in. It was a slow job which had to be done carefully. The level of destruction was so great that in some cases it was possible that the only remains of some people would be a tooth, a nail or even a hair.[56] Imagine trying to find this among the carnage of the plane wreckage with so many people walking across the site, contaminating it in the process. The search process became worse day by day. Despite being some 1,500 m above sea level, temperatures were in excess of 30 °C. On some days there was light rain. Flies and bugs were everywhere. Maggots began to infest the remains. The stench of dead bodies hung in the air – to the extent that many who travelled to the site found it difficult to wash the smell away at the end of a visit. Remains were hanging from trees as well as being on the ground. One soldier thought he had found a monkey which had been killed as the plane crashed on the mountainside, but subsequently realized that it was the unborn foetus of one of the victims.[57] The death toll is therefore considered by some to be 521 rather than 520. Working in these conditions, it is no wonder that there are reports of soldiers subsequently being sick, suffering from post-traumatic stress disorder (PTSD) and not being able to eat.[58]

As the remains were brought to Fujioka they were taken to the Shimin Kaikan (civic hall), which had been set up as a temporary morgue. However, as the number of remains increased, additional locations were needed, so two high school gymnasiums were also used.[59] Here, doctors and dentists conducted the identification process, which continued until 18 December, a total of 117 days.[60] There were so many doctors and dentists at one point that one *izoku* reflected that he felt that there were 'more dentists than victims'.[61] They all volunteered their time to do this, most were from within Gunma prefecture and included Tsutomu Ōkuni, noted for his work on the infamous Ōkubo case.[g] The conditions in which they worked were harsh. Windows had to be closed with sheets placed across them so that the media could not see inside. The temperature was estimated to be as high as 40 °C. Incense was burnt to help mask the smell of the remains. Some of the doctors kept working with little, or no sleep. Those that did sleep sometimes rested on the floor or desks rather than leaving the hall.[62] There were also nurses present. One of their roles was to clean and prepare the remains prior to *izoku* being asked to make identification. The intention appears to have been to remove blood stains, for example, so as not to shock the *izoku* and to help with the work being done by the doctors.[63] Additionally, for those where the physical damage was greater, bandages were used to help reform and reshape the face, again, in order to help reduce the mental torment for the *izoku*, but also to aid in the identification process so that the victim was more recognizable. Although some *izoku* have criticized this process for divorcing them from fully understanding the scale of what their loved one(s) experienced,[64] it would appear that most were appreciative of these efforts.[65]

The main role of the doctors and dentists was to categorize the remains which they had in front of them. No autopsies were conducted. The cause of death was written as *Nikkōki Tsuiraku* (JAL aircraft accident) for all, with the time of death being recorded as 'about 7pm, 12 August, *Shōwa* 60 [1985]'. The lack of autopsies was a major failing. Autopsies might have clarified which victims died as a result of initial crash and which victims died subsequently during the night. Not only would this have highlighted the impact of the tardy SAR, it could also have had implications for aircraft design. As Kawakita points out, and the organization, IREI Air-Safety, which he co-founded has continued to promote, knowledge about the effectiveness of seat belt designs, designs of seats and overhead storage compartments is a critical element in trying to make more crashes survivable.[66] Some on JL123 were seemingly killed by contents coming from overhead compartments. Others were sliced in half by the seat belts which were supposed to help them. If more was known about the number and the process, it may be possible to better design planes to reduce such risks.[67] The lack of autopsies erased the possibility of collecting this valuable data. As mentioned in Chapter 4, the lack of autopsies, and therefore the inability to include information from them in the official accident report, meant that information such as the captain's teeth being ground down were not taken into account by the investigation into the crash. Such small pieces of information can provide useful insights into what took place. Even now, autopsies are carried out on only about ten per cent of suspicious deaths, and there are many who feel that not enough autopsies are conducted in Japan.[68] But clearly for some Japanese there are reasons for autopsies not to be done, such as showing respect to the body, in which some believe the spirit remains for a period of time after death (see Chapter 6).

When doctors or dentists had completed their investigation, an information sheet, as shown in Figure 5.6, was placed on the coffin to aid the *izoku* attempting to identify their loved one(s). After the sheets had been placed on coffins, *izoku* were able to enter the makeshift morgues to attempt to identify their loved one(s) (see Figure 5.7). Initially, there was a restriction of one representative from each family due to the lack of space, and it could take as long as three days to go round the morgue once.[69] However, this restriction was eventually lifted. Issues relating to the symbolism of how the coffins were presented will be discussed in Chapter 6.

Figure 5.8 shows the days on which bodies were identified as well as the number of complete bodies and body parts which were retrieved on those days from the crash site. There are three issues which are highlighted by this figure. First, the total number of victims identified did not reach the total number of victims. In the end, 518 of the 520 JL123 victims were identified. Second, one can see the huge number of body parts retrieved from the crash site. As each one of these had to initially be stored separately in its own coffin, a total of 2,086 coffins were needed.[70] The number was so great that coffins had to be brought in from

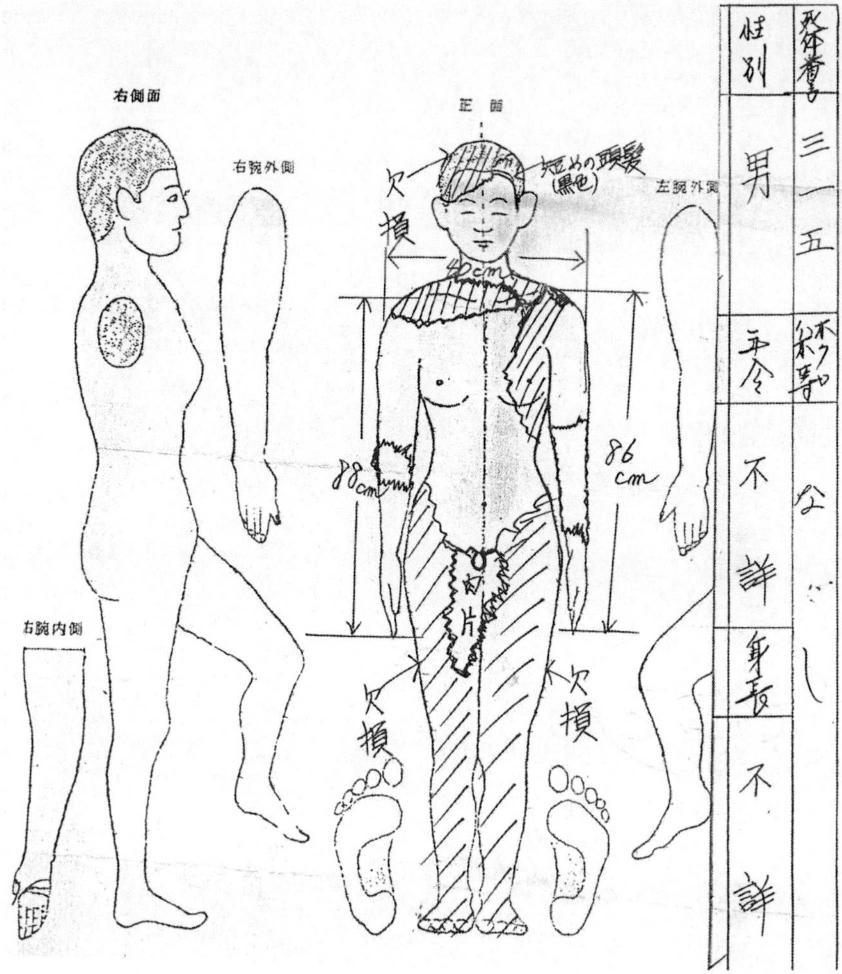

Figure 5.6 Information sheet placed on coffins for identification

Picture courtesy of Peter Mathews. This information sheet was for coffin number 315. The victim was male, but the age and overall height was undetermined. The diagrams provide details about some of the victim's measurements and which parts are missing from the remains.

outside the prefecture, as were hearses to carry the coffins, and taxis to ferry *izoku* around.[71] So as to maintain the remains as best as possible, the coffins were filled with dry ice. Each coffin needed at least three 2.5 kg blocks in it. In total 110,000 blocks of dry ice were needed, costing about ¥82m.[72] Despite this, partly due to the time taken to find body parts at the crash site, many remains became infested with bugs and maggots.[73] Consequently, there are reports of people not being able to eat rice due to the visual reminder of this.[74] Similarly as many of the remains were burnt, some doctors since found it hard to eat grilled chicken.[75] The third

Figure 5.7 Inside the temporary morgue

Picture courtesy of Peter Mathews.

point that the graph shows is the apparently large number of 'complete bodies' which were retrieved from the crash site. However, a 'complete body' does not necessary mean that the whole body was found intact or that all parts of the body were connected. The critical elements in defining a body as complete is whether the five parts (or some part thereof) of the head, two arms and two legs are present. In addition, so long as just one part of the head was attached to the body, it was included as a 'complete body'.[76] In fact, only 177 of the 'complete bodies' had all the parts of the body connected.[77]

Hiroshi Makizono 牧園弘志 *(37)* · *Isao Masuda* 増田勇生 *(26)* · *Kazuhiko Masuda* 益田和彦 *(49)*

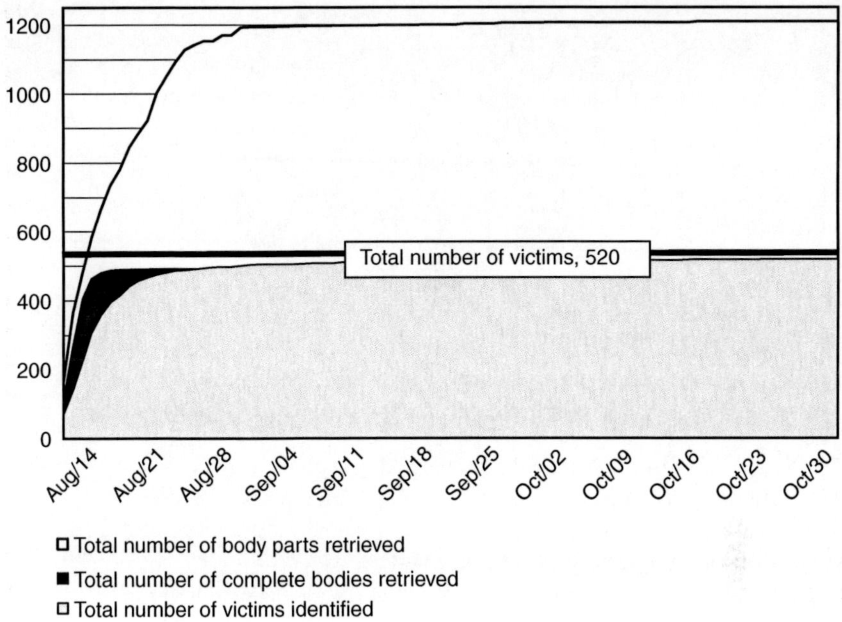

Figure 5.8 Identifying the remains

Figure by the author, based on data in Gunma-ken Sōmubu, 1986:48–53 and Kawamura 2005:114–15.

A large number of remains were never identified. The emphasis appeared to have been on identifying that a loved one had been on JL123 rather than ensuring that all of their remains were identified. As will be discussed in Chapter 6, certain anniversaries after someone dies are significant, and being able to identify and conduct a funeral in time for these is important. On top of this, with many families waiting in the school gyms in Fujioka, there appears to have been pressure to try to speed up the identification process. But the process was consequently flawed. As remains started being taken away for funerals, there was nothing for other remains to be compared with. As *izoku* left Fujioka with some remains, other remains belonging to their loved one stayed in Fujioka never to be identified if they did not return again. Although the situation may be understandable, it was perhaps not the appropriate way to proceed. It needed someone to take control and explain that the identification process needed to be done properly and effectively and that it would be a time-consuming process. Although some *izoku* may have objected, there would have been others who understand. The extra time would have probably reduced the concern that errors in identification – where an *izoku* might have incorrectly identified remains as their loved one's – could have been made. Kawakita points out the way in which the process was done following the

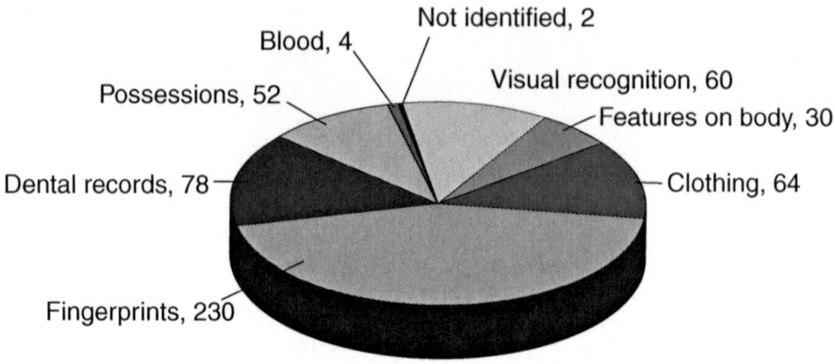

Figure 5.9 How remains were identified

Figure by the author, based on data in Iizuka 2001:174.

Air New Zealand Flight NZ901 crash on Mount Erebus in 1978 when families were told that they would not get the remains of their loved ones returned to them for several days or weeks as time was needed to conduct autopsies and ensure that the identification process was complete and accurate.[78]

Let us now turn to looking at the way in which identification of the 518 identified victims was conducted. Figure 5.9 summarizes this. As some bodies were relatively undamaged, *mensetsu* (which is usually translated as 'interview' but here would mean 'visual recognition') and personal effects were used in many cases for identification. Neither of these methods is generally recommended, as mistakes can be made by those doing the indentifying,[79] or incorrect belongings could have ended up being placed with the corpse.[80] As a large number of bodies were badly burnt or incomplete, other methods had to be used. Dental records were very important, despite the inherent problems in this identification method as it can be hard to match up existing records with any photographs of the victim since variations in the angle of the camera can cause differences which mean that the two images do not look similar.[81] Nearly half of the victims were identified by their fingerprints, although it was noted that in some cases it was hard for the *izoku* to find a fingerprint for comparison due to cleaning up the house after hearing the news about the crash.[h]

With the crash site being so large, Gunma Police developed a grid of the site to make understanding the location easier. Once the remains had been identified in Fujioka it was possible to match this with the information gathered at the crash site. Later, metal posts were placed in the appropriate position to mark approximately where each identified victim was found. What has happened subsequently will be discussed in Chapter 6. Using the information about where each victim was found, it is possible to map out where passengers from various parts of the plane were found (see Figure 5.10). This helps give a better understanding of the way in which the plane broke up on impact with the ridge. Naturally, the location of where victims were sitting and where their remains were located also has a correlation with how they were identified. Those towards the rear were

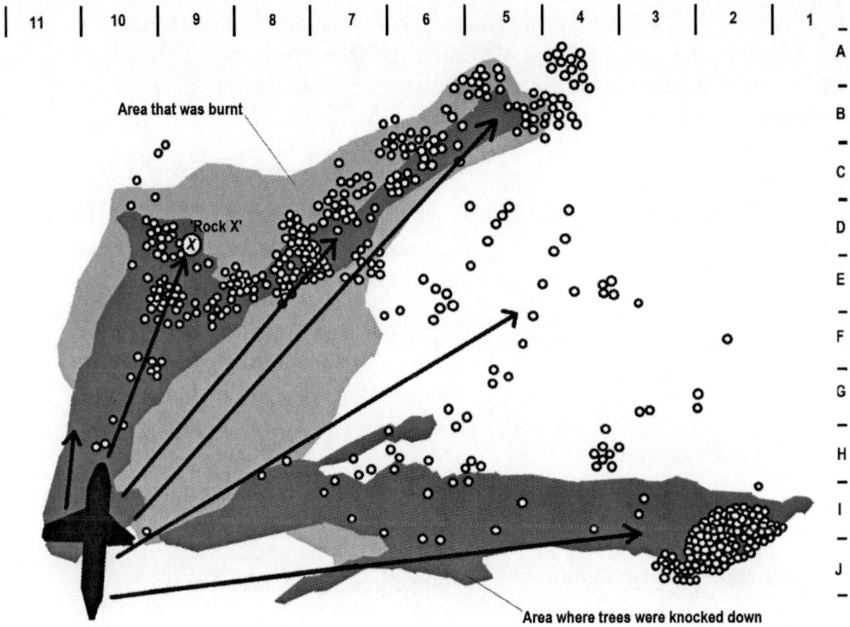

Figure 5.10 Osutaka-no-One: where remains were found

Figure by the author, based on Kawamura 2005:56; Iizuka 2001:80; Aoyama 2010:402.

generally 'complete bodies' and could be largely identified through *mensetsu* or using information from belongings on their bodies. The bodies of many passengers from the middle section of the plane were either incomplete or had been burnt beyond recognition having been sitting in the area above the fuel tanks, but identification using of fingerprints and dental records was possible. Those in the front of the plane, having experienced force in excess of 100 G, were largely only identifiable through a few fragments of bone.

Identifying personal effects

There were also the passengers' *ihin* (personal effects) to be identified. Each was catalogued and *izoku* then had the opportunity to try to identify them. This was not a straightforward process either. To have a watch returned, for example, the *izoku* had to produce a record of the serial number or receipt.[82] Amongst the *ihin* there are a few particular items which are worthy of additional discussion.

The camera belonging to Tetsu Ogawa, which contained some pictures taken during the flight, is discussed elsewhere. This was not the only camera to have survived the flight,[i] though seemingly only Mr Ogawa had pictures taken on JL123. Police were careful to check the contents of tapes on all recordable personal stereos in case there was any information there which would be useful to the investigation into the cause of the crash.[83] There was another reason to check the tapes. A few days after the crash a number of *isho* were found on passengers' remains, although sometimes not until their belongings were checked at home.[84] Ordinarily, *isho* would be translated as 'a will' or 'a testament', but this does not quite fit here. One of the *isho* found was written by one of the stewardesses, Yumiko Tsushima, and was a list of what needed to be said, in both Japanese and English, to passengers following an emergency landing. Whether written to ensure she did not forget anything, or as a means to distract her attention from the inevitability of what was likely to happen next, will never be known, but one cannot help but be moved by the apparent high level of professionalism that was demonstrated in such a severe situation.

Yumiko Tsushima

(Underlined parts indicate those which were originally written in English.)

Stay calm
Unfasten your safety belts. Brace yourselves. Don't take any baggage
Follow our instructions
First announcement to PAX[j]
Check if the doors can be used
C'K for fire outside. C'K with other CREW
Unfasten your safety belts. High heels
Don't take any baggage
Persons in front 2 rows
Jump. Jump and slide
Get away from the aircraft
Go to a safe area
Take off high heel shoes. Don't take any baggage
Help the elderly and disabled
Release your seat belts
Remove high[k]
Don't take baggage
Follow our instruction
Fire
Lower yourself and cover your mouth and nose with towel
Follow the person in front and move over there
Low position with a wet towel
Covering nose and mouth

As will be discussed in Chapter 8, there are many things for which JL123 is remembered, but the *isho* is certainly one of them. The updated version of the book on disasters which I had read as a child (see Chapter 1), names its section about JL123 ‘Love letters in the sky’ because of the *isho*.[85] Sometimes one gets the impression that there are those who believe that everyone, or certainly many, wrote *isho* on JL123. I have even come across references to Kyū Sakamoto having written an *isho*.[1] This does not seem to have happened. Nor is it the case that many on JL123 wrote *isho*. Only seven *isho* were recovered. Although others may have been written that were subsequently destroyed in the fire, it is an exaggeration to think that passengers and crew alike spent their last 32 minutes frantically writing messages to loved ones. If we look at where the six passengers who wrote *isho* were seated (see Figure 3.5), we see they were sitting in seats 20H, 22H, 29B, 48C, 48D and 53C. It is quite plausible that one of those in 48C or 48D saw what the other was doing and was inspired to do the same. But otherwise, the gaps between these six passengers would suggest that, unless many other *isho* were written and were not recovered, they were the result of independent actions by a small minority of the passengers. Despite this, they are still worthy of study in trying to understand Japanese responses to such a situation, since all of them were Japanese. Merely studying what the majority does is one-sided, biased and tends to do little other than reinforce stereotypes. Of course, studying exceptions can exaggerate their significance, but as the *isho* provide an insight into what at least *some* were feeling, as well as them being such a well-known feature of the JL123 story, I believe it warrants reproducing all those written by the passengers. All six are, of course, translations. I have used the translations on display at the JAL Safety Promotion Center as a basis, but have made a few alterations to try to preserve the nuance of the original Japanese. Annotations are provided in footnotes to help explain some of the choices of words. What is lost in reproducing the words here – other than the example at the start of this chapter – is the impact due to the nature of how it was written, what it was written on, or other visual factors (such as staining of blood or earth from the crash site). I was fortunate enough to be able to see the actual Kawaguchi *isho* for myself. Touching and seeing the pages of the diary, which includes a reminder of being on JL123 on 12 August 1985, was very emotional and is an issue I will return to in Chapter 9.

Keiichi Matsumoto

PM 6:30
Tomoko
Look after Tetsuya (and parents)
Keiichi
Suddenly there was an explosion and the masks dropped
With the explosion we began to fall
Be brave and live
Tetsuya be good

Shinjirō Minami 南慎二郎 *(54)* · *Kazuyuki Minoda* 箕田一行 *(35)* · *Hiroshi Misaki* 三崎洋 *(45)*

Ryōhei Murakami

The plane is swaying a lot left and right,
18:30 descending rapidly
Flying steady
Japan Air Lines 18:00 flight to Ōsaka accident
I might die.
Murakami Ryōhei
Everybody please live happily.
Goodbye Sumiko Miki Kyoko Kentarō
18:45 the plane is level and stable
There's little oxygen, I feel sick
Inside the plane voices are saying let's do our best[m]
I don't know what happened to the plane
18:46 I am worried about the landing
The stewardesses are calm

Mariko Shirai[n]

I'm scared. I'm scared. I'm scared. Help me. I feel sick. I don't want to die. Mariko
Keiji, Hisako, Tadaomi, Shin'ichi, Rihiya, Sakura

Masakazu Taniguchi

Machiko,
Look after the children
Ōsaka Minō
Taniguchi Masakazu
6 30

Kazuo Yoshimura

Please live bravely. Please look after the children

Walter has noted that one question which many families of war dead want to know the answer to was 'what were their last words?'[86] These days we rarely hear about the last words of people, whereas the last words of kings, queens and many others across the centuries have become as well known as their achievements. The *isho* were not the last words, but they are a substitute for that. That 'last words' are both of interest and are rarely mentioned in relation to those who die in the modern era is perhaps one reason why the JL123 *isho* generate such interest. Of course, that they were written in unusual circumstances is another factor why they, and this flight, are remembered. Most air accidents do not present an opportunity for such notes to be written. The time between the initial accident and the

Ken Miyajima 美谷島健 *(9)* · *Masayuki Miyane* 宮根将行 *(45)* · *Seiichi Miyaoku* 宮奥誠一 *(52)*

eventual crash is usually too brief. The phone messages and calls made by those on the 9/11 flights are comparable, but a comparison between these messages and the JL123 *isho* is beyond the scope of this study.

We have to keep in mind the *isho*'s impact on the *izoku*. I can imagine that the desire to fulfil wishes made in the *isho* could be a burden to families. Similarly, whereas many *izoku* have the hope that their loved ones were comforting each other if travelling in a group, or were being comforted by another passenger if travelling alone, Mariko Shirai's *isho* probably erases all such hope. I have also spoken to *izoku* who have wondered why their loved one did not write them an *isho* and this produces yet another burden.

If we consider the contents of the passenger *isho* themselves, four directly mention the time when it was first written. The starting times of these four are about the same time at 18:30 (allowing for slight differences in individual's watches and a tendency for some to round to the nearest five minutes). One has to assume that they decided to write the messages after the initial commotion caused by the explosion and need to put on oxygen masks had calmed down. Of course, there is no way that at this time they could have known that they only had a further 26 minutes to live, but their messages certainly suggest an awareness that they were unlikely to live for much longer. However, only one of these four provided further updates about the plane's situation and how he was feeling. Having taken the decision to write something, why did most not write more? Sometimes the lack of space was the answer. But even within the space available they do not particularly ramble, as though the extremity of the situation produced a clarity about what they wanted to be expressed.

A feature which some of the *isho* have in common is the use of the word '*shikkari*'. Dictionary suggestions for this word include 'firmly, tightly, reliable, level-headed and steady'. For the JL123 *isho*, it has either been translated as 'brave' or left out. I do not think these words really match the nuance of its usage in the *isho*. Indeed, the *isho* do not translate well into English. None of the *isho* include the word 'love', but whilst the word is not used, the messages are not devoid of love, it is just not expressed in the same way in Japanese as it is in English. Many of the *isho* were written by fathers writing to their families and this was their 'spirit's cry'.[87] Ochiai's interviews with Yoshioka also suggests that the men appeared to not be coping with the situation as well as the women,[88] although Hiroko Yoshizaki points out that her husband's advice to remove her glasses may have helped save her life and showed his awareness of the situation and what to do.[89] It is interesting that one *isho* points to a certain togetherness among the passengers with the phrase 'voices are saying let's do our best'. Although Ochiai does not mention this particular event, her interview also suggests that there was no panic.[90]

Another feature of some of the *isho* is mention of what was happening to the plane itself. Some of this information is significant, but, as with the lack of photographs of the damage taken by Mr Ogawa (see Chapter 4), some of what is not mentioned may be of greater significance. One *isho* mentions 'smoke', which is likely to be referring to the mist produced when the decompression happened, but no mention is given to ears popping or hurting or air being sucked from inside them. There is no mention of a breeze or wind. There is no mention of the plane becoming cold. *Isho* mention the plane descending rapidly at 18:30 – but the FDR does not show this. Is this merely that a passenger's idea of what 'descending rapidly' means and what a steep descent really is are not the same?

Compensation

Both this chapter and Chapter 4 have discussed responsibility for the crash and the way in which the *izoku* were looked after. These two areas come together in the form of compensation. Compensation is different to an insurance payment in that it is paid out by a party or parties on their own accord or following a legal ruling to do so. Insurance payments are only made if the individual has a policy.[o] Of course the insurance world is also involved with compensation in that the party or parties making compensation payments will seek to reclaim money from their insurance. This is likely to increase further insurance premiums or potentially make it harder for them to get insurance coverage. The higher the amount paid out, the greater that problem becomes. Consequently there is likely to be a desire to keep compensation payments as low as possible. When dealing with 520 fatalities, it was clear that total compensation payments were going to be extremely high.

Due to religious 'beliefs' (see Chapter 6), JAL agreed that discussions about compensation would not begin until forty-nine days after the accident. With a large number of funerals being conducted prior to this date, JAL made payments to all families resident in Japan of ¥500,000 as *omimaikin* (sympathy money), ¥500,000 as *kōden* (funeral offering to the dead) and ¥500,000 as *sōgiryōkin* (funeral expenses).[91] None of this is regarded – by the company or those receiving it – as any sign of the company admitting liability for the accident, but due to the legal 'difficulties in applying this Japanese custom', it was not offered to those *izoku* outside of Japan.[92] It was subsequently agreed between JAL and Boeing that Boeing would pay 80 per cent of the compensation and JAL the remaining 20 per cent, and JAL would act as the contact point.[93] Given that the probable cause is said to have been due to the faulty repair work done by Boeing and that this was something that Boeing itself admitted to, it is perhaps a little surprising that Boeing did not end up having to cover 100 per cent of the compensation.

The exact details of the compensation received is not in the public domain. However, some of the details are known; an article in the *Sankei Shimbun* pointed out that whilst there was a limit of ¥2.6m for international flights, 'the sky is the limit' for domestic flights and suggested that sums of ¥30m to ¥99m were paid in the case of the JAL crash in 1982.[94] Yoshioka suggests figures of ¥60m and ¥70m in the case of JL123, making the total between ¥31.2bn and ¥36.4bn.[95] But that the process in getting compensation was far from straightforward is highlighted by the fact that it was one of the main issues about which the 8/12 Renrakukai was most active in its early stages (see Chapter 6). Negotiations essentially happened at an individual family level. JAL also had scholarships for children's education which merely varied with the age of the child when they were in formal education.[96] Some of the problems appeared to centre around there being two types of calculation being used for the main compensation.[97] The key issue for some families was that as calculations were based on 'normal assumptions' about Japanese society at that time, it assumed a lower expected income for women than men. Although this could clearly be supported by statistical data, for the *izoku* this was missing the main point; the compensation was a payment because their loved one having died. On top of this, some *izoku* felt that the way in which their loved one had died warranted additional consideration.[98] This was not a view felt by all *izoku*. One *izoku* I spoke to, perhaps because of their medical background, finds it hard to comprehend this emotional element added to death, which they just see as a medical fact.[99] This difference of view highlights one of the puzzling elements of compensation. Who receives compensation, why do they receive it and how much should they receive?

Compensation appears to have become a way in which societies driven by money have come to deal with emotions. But because emotions are subjective, amounts of compensation will vary. Compensation cannot solve emotional issues. In a more financially and litigiously driven world, there appears to be an expectation among many that such monies should be forthcoming. Although the accident resulted in no deaths, some who were on British Airways flight BA038, a Boeing 777, which crashed at Heathrow Airport in January 2008, sought to sue Boeing, due to the mental trauma they suffered.[100] The idea that 'accidents happen' is being replaced by one that those who cause accidents should be made to pay for it. However, the cost of such compensation is additional to any payments needed to ensure that similar things do not occur again. In some cases this has the potential to be financially crippling. There is a danger in such situations that those who could be part of a system to ensure that lessons are learnt, will instead end up in positions where this knowledge is not being utilized. This argument is similar to the one raised in Chapter 4 about responsibility and the appropriate form of punishment.

In considering the amount of compensation, whatever the source, one of the key factors appears to be the nature of the death. Mass deaths of people of the right background appear to lead to greater amounts of compensation. There is an expectation that the company or companies responsible will pay out large compensation following a plane crash. A disaster at a football stadium, on the other hand, may see the balance shift more to public donations. Key elements appear to be the degree to which a large number of people were involved and that those involved were killed due to something beyond their control. That the way in which they died may have been particularly traumatic could further increase the desire to compensate. There being children among the victims is also likely to exaggerate the response. A car crash is unlikely to see any compensation unless it was due to a manufacturer's design fault and insurance will be the means of providing financial support. Victims (or their surviving relatives) of cancer, for example, will see little compensation. Again, the individual's own health insurance, if they had any, will be key. There is not likely to be any compensation or

Mami Morita 森田麻美 (22) · Giancarlo Moroni (39) · Andrea Moroni (17)

donations from the public. In comparison to those who die in a plane crash, they have died the wrong death in the wrong circumstances.

In Japan there have been a number of cases of groups struggling to get compensation. For example, there have been non-Japanese victims of Japan's actions in Asia during the war, Japanese victims of the war (such as the fire bombing of Tōkyō in March 1945 or the bombings of Hiroshima and Nagasaki), the victims of Minamata Disease, and victims of a scandal which saw haemophiliacs receive HIV positive blood. Many of these cases have been largely unsuccessful and/or have been very lengthy procedures. They were also largely seeking compensation from the state in some form. By comparison, compensation relating to JL123 was completed much faster. More than 90 per cent of *izoku* settled with JAL without going to court,[101] but this is not to say that there were no problems. Indeed, the final compensation cases relating to JL123 were not completed until 1995.[102] Or at least, this is how it seemed until the case of Cassie and Diana Yukawa, whose father died on JL123. Compensation went to the Yukawa family, but as their mother was not married to Mr Yukawa, they did not receive direct compensation, although money came from Mr Yukawa's mother. When she died, steps were taken to get the girls additional compensation. Although JAL was 'surprised' the application came so late, and that the amount was in addition to what had been agreed with the Yukawa family,[103] a form of compensation was agreed upon.[104] This compensation allowed Diana to receive violin tuition from some of the finest violin instructors in the world and for her elder sister Cassie to also be able to complete her music studies.[105] Diana has gone on to become a renowned violinist and featured on the soundtrack for *Shizumanu Taiyō* (see Chapter 8).

Conclusion

This chapter has covered a variety of different areas relating to the human cost of the accident; how the remains were gathered; how they were identified; issues relating to the belongings (including *isho*); and the compensation paid to families. The chapter has highlighted how many different groups of people were involved, what these people did, and some of what they experienced. The chapter has pointed out that Japan's civil society was alive and well in 1985, before the supposed '*gannen*' of 1995. This aspect of the JL123 story will be developed further in the following chapter. However, in identifying the great efforts made by the various municipalities in Gunma in responding to the crash, it is worth concluding this chapter with a summary of the financial costs they faced, and how these were recuperated (see Table 5.2).

Table 5.2 The cost of the crash to Gunma

Area		*Total Cost (¥ million)*	*Relative Burden*	
			JAL (¥ million/%)	*Government (¥ million/%)*
Gunma Prefecture		762	455 (59.7%)	307 (40.3%)
Municipalities	Ueno-mura	32	26 (81.3%)	6 (18.8%)
	Fujioka	21	18 (85.7%)	3 (14.3%)
	Ten other municipalities and three associations	21	16 (76.2%)	5 (23.8%)
	Sub-total	74	60 (81.1%)	14 (18.9%)
Total		836	515 (61.6%)	321 (38.4%)

Table by the author.
Source: Gunma-ken Sōmubu 1986:33

6 Remembrance and the Osutaka Pilgrimage

Look up at the stars at night,
The tiny light of a tiny star is singing about a meagre happiness.
Look up at the stars at night,
Just like us, a nameless star is praying for a meagre happiness.[a]

Having looked at issues relating to how the victims of JL123 were identified and how the *izoku* were looked after and compensated, this chapter looks at how the *izoku* have continued to work to ensure that JL123 is never forgotten. The first section of this chapter will look at the way in which some of the *izoku* came together to form an association which has been one of the main driving forces in the remembrance activities. The chapter then goes on to consider Japanese attitudes to death. Although we have discussed various issues relating to death in previous chapters, many of the key issues relate to remembrance, which is the theme for this chapter. This section will also include discussion about Japanese attitudes to 'religion' and what this term means. The final section will then look at how the victims of JL123 are remembered in what I call the Osutaka Pilgrimage.

8/12 Renrakukai

On 7 December 1985, 15 *izoku* came together at Kamata Seikatsu Sentā to discuss forming a group, which then met in Maebashi after the mass cremation on 20 December.[1] Concerns relating to compensation, the investigation into the cause of the crash and the various mental and other problems some were suffering were discussed.[2] As a result of the meeting, the *izoku* formed an association, headed by Kuniko Miyajima (see Chapter 7), called the '8·12 Renrakukai' (8/12 contact association).[b] One thing that is striking about the name of the group is that the word *izoku* is not used. The first meeting made it clear that they did not want to use this term as it was overly negative and would not reflect the association's purpose.[3]

The main aims of the 8/12 Renrakukai are to 'comfort the spirits' (*irei*), research the cause of the crash and promote future air safety, and to provide assistance to its members.[4] In their early years, the activities included collecting signatures of

Hideto Nagai 長井秀人 (26) · Akio Nagaoka 長岡明男 (24) · Masamichi Nagaoka 長岡正通 (37)

the public in the streets in support of their association and calls for a reinvestigation into the cause of the crash and that criminal proceedings be brought against those responsible.[5] Although this has been the most public aspect of the 8/12 Renrakukai at one level, much of the work has been away from the public eye. The association is a support network. Members can contact each other to discuss their feelings and problems, and to aid this process a regular newsletter, *Osutaka*, is produced. Additionally each August an additional publication, *Akanegumo* (madder red clouds),[c] is produced. *Osutaka* contains the latest information about what the association is doing, letters from members, key developments relating to JL123, JAL and related issues. It also contains copies of newspaper articles relating to these issues. To date the letters which appear in *Akanegumo* have been published as three books. The third of these, published to coincide with the twentieth anniversary, points out some of the statistics relating to the *izoku*: the 520 victims came from 401 different families; 189 families were left with no husband/father, 22 families lost both parents and all the children, seven widows went on to have children after the crash.[6] The problems that some of these *izoku* were facing was not helped by the language, whereby one of the words for widow (*mibōjin*) literally means 'not yet dead person'.[7] There were at least three cases where families separated or divorced subsequent to the crash.[8]

As well as providing a means of support for members, the 8/12 Renrakukai effectively provides a means of communication between JAL and *izoku*. However, this may, in some cases, be at the expense of effective communication between JAL and *all izoku* as not all *izoku* are members of the 8/12 Renrakukai. The *izoku* of the 22 foreigners are not members, for example. Indeed, all 8/12 Renrakukai communications are done in Japanese. Although all *izoku* have a contact person within JAL, it would seem that this person does not always pass on the same levels of information as the 8/12 Renrakukai does to its members.[9] A prime example of this is in relation to the anniversary events held each year in Ueno-mura. The 8/12 Renrakukai has become so much the co-ordinating body for these that non-members, particularly foreign non-members, are seemingly not involved in the discussions and may not be fully informed of what is planned.[10] This would suggest that the system is not perfect, but it also means that it is important not to suggest that the 8/12 Renrakukai as being a united group of all *izoku*. Even within the 8/12 Renrakukai there is a variety of different views and ideas about how they should operate. One of the consequences of this would appear to be the creation of *Kōkū Anzen Kokusai Rarī* (the International Rally of Endeavors to Improve Air Safety, or IREI Air Safety) by some of the *izoku,* including members of the 8/12 Renrakukai. IREI Air Safety is much more focused upon discussions relating to the cause of the crash and improving air safety,[11] although it has also been involved in certain remembrance activities (see below). These issues of unity are relevant in relation to discussions of 'communitas' when looking at the pilgrimage made to Ueno-mura (see below).

The 8/12 Renrakukai also provides a natural link between many of the *izoku* and the volunteers from Gunma (see Chapter 5). Many of these volunteers continue to be involved in the anniversary events and also arrange a number of other events during the year. Of course, the 8/12 Renrakukai is a voluntary group, and further reflects the fact that Japan has an active civil society (see Chapter 5). Due to the way in which 8/12 Renrakukai developed, it has become a beacon for *izoku* of other transport-related tragedies and they are often involved in helping such *izoku* form their own association and providing support and advice (see below).

Death and religion in Japan

Death is a recurrent theme throughout this book. That so many died on JL123 is one the reason why it is of such interest today (see Chapter 8). Yet, partly as there were so many, the victims are often reduced to a mere statistic and the consequences of such an accident are rarely covered – particularly in the English-language sources. To appreciate some of the issues discussed throughout this book, we need to have an understanding of Japanese attitudes to death and, linked to this, to religion. It is important to emphasize the ‘s’ in ‘attitudes’ in the previous sentence, as I do not want to suggest that all Japanese share a common attitude about death or religion.

Most Japanese are both ‘believers’ in Shintō and Buddhism. Official statistics suggest there are over 106 million of the former and 95 million of the latter, although the population of Japan is around 127 million.[12] But rather than ‘belief’, it is practice that tends to be of significance, which is why many Japanese have no issue practising (or believing) in both. This also helps to explain why between 70 and 80 per cent of Japanese say that they do not believe in a religion.[13] The concept of *shūkyō* (religion) tends to be a ‘narrowly based, theologically and doctrinally orientated phenomenon, associated with belief in a specific faith’,[14] and this does not necessarily fit with how Japanese people behave. In this respect, religion includes the

> observance of calendrical events and festivals, ranging from memorial visits to graves to New Year visits to shrines and temples, and with matters such as the purchase of amulets and other lucky charms, visits to diviners, attitudes toward spirits, and the potential incidence of miracles.[15]

Naturally, practice is based on certain ideas and philosophies contained within Shintō and Buddhism, together with elements of Confucianism, but there is not the same level of doctrine found in many other world religions. With the separation of education and religion, much is learnt through behaviour. There is no requirement to go to a temple or shrine on a particular day of the week, for example, so no opportunity for a priest to reinforce certain philosophies or moral codes. Attendance at temples and shrines is connected to certain events and moments in a person’s life.[16] One consequence of this is that Japanese institutions do not discriminate to whom they provide services and materials. As Reader and Tanabe point out,

> Anyone may purchase an amulet, request a special ritual prayer, or make a personal supplication to the gods or buddhas… No one is barred from

searching for good fortune, seeking the providence of Kannon, imbibing the holy powers, and receiving the benefits there.[17]

By extension, the deities (whether Shintō or Buddhist) 'are there, in the popular view, for the benefit of humans, to be called on when someone is in need, to help and support them'.[18]

When it comes to death, people will generally turn to Buddhism, as for Shintō, death is dirty (*kegare*) and anything and anybody associated with death should be avoided. Indeed, funeral and memorial rites are the 'the single most important religious function of Buddhist priests and temples'.[19] However, Buddhism and Shintō do come together in relation to ancestral worship, perhaps among the most important of all religious activities in Japan,[20] although it is still Buddhist practices that dominate. But I would suggest that many of these practices are based upon doing what is expected rather than necessarily doing it because of an underlying understanding or belief in what is being done. For example, two of the significant Buddhist rituals relating to death take place on the seventh and forty-ninth day after an individual's death. But I have spoken to *izoku* who do not necessarily believe in the significance of these days. However, Kuniko Miyajima, for example, believes that her son's spirit entered her some time later and so he is always with her. In that respect there is belief that there is an existence beyond physical death.[21]

The important point in relation to these two ceremonies is that they are based on a belief that the individual is not fully dead when they are physically dead. The ceremonies are concerned with ensuring the individual's spirit makes its way safely to the other side and that they become *hotoke* ('a Buddha' or 'the dead').[22] Due to the significance of this JAL was careful not begin discussion about compensation until after 49 days had passed.[23] Further remembrance activities are required in subsequent years so that the individual may then become an ancestor, which does not happen until the thirty-third or fiftieth anniversary depending on the Buddhist sect.[d] It is important to understand that when the individual becomes an ancestor that they effectively become part of a nebulous entity and that their individuality is lost. After death many will gain a Buddhist posthumous name (*kaimyō* or *hōmyō*), but even this is not used once they become an ancestor.[e] Unlike in a family tree, which shows connections to named individuals, Japanese ancestors are a nameless entity which is thought to look over the living, and so a potential source of benefits.[24] I would suggest that this can be best understood by considering a diagram of a hot balloon (see Figure 6.1). The basket contains the living, who are known, named individuals. The part above this is the recently deceased. This includes, at the lower level, those who are physically dead, but not yet considered to be Buddhas, and at the upper level those who are both physically and spiritually dead. These are known individuals referred to by either their original name or a Buddhist posthumous name. The main part of the balloon is made up of all the ancestors. They have no individuality and the names are no longer important.

One of the significant factors in relation to the deaths of those on JL123 was that they were 'untimely'. Although we all know that one day we will die, the

Minoru Naitou 内藤実 *(33)* · *Hiroshi Nakabeppu* 中別府弘 *(48)* · *Kayoko Nakada* 中田加代子 *(38)*

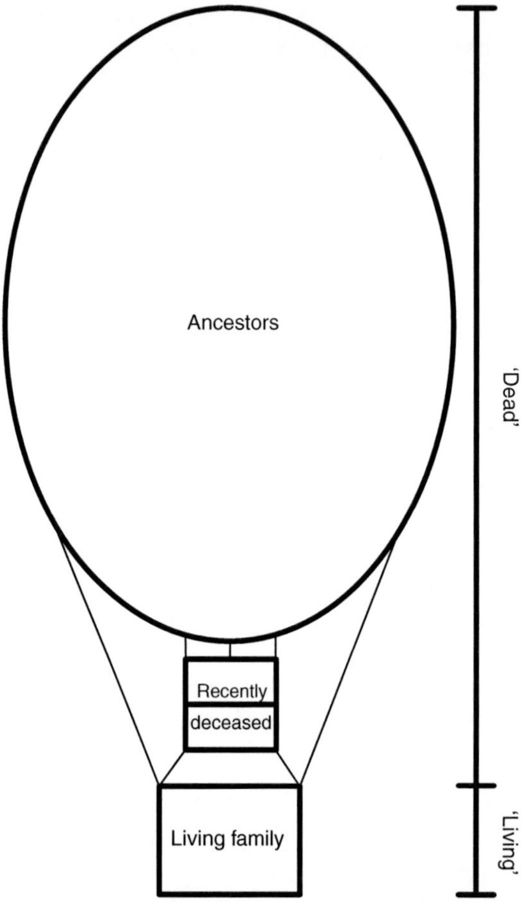

Figure 6.1 Ancestor worship

Figure by the author.

timing of our death is largely unknown. For many it will follow a period of ill health, which becomes a time for both the individual and those that know them to reflect and prepare. For those to be left behind, it allows for a degree of preparation for life without their loved one, to prepare for a future which can reflect on the past of a loved one who has no present. But preparation requires time and some lives are 'interrupted',[25] and suffer an 'untimely'[26] or 'abnormal'[27] death. In other words, they do not die due to 'natural causes' at the end of a 'natural' lifecycle, what could be termed a 'normal death' (*seijōshi)*.[28] There is little or no preparation time. Also, as van Bremen points out, 'An "untimely death" is also a "bad death"' as it 'produces the most unsettled spirits that need to be placated in order to remove their emitting dangerous and deadly retribution (*tatari)'*,[29] as discussed below.

Mika Nakada 中田美加 (12) · Mineko Nakagami 中上岑子 (38) · Yoshiya Nakagami 中上義哉 (12)

'Untimely' deaths are statistically rare, but this rarity may be one of the factors that drives the fascination with them amongst the public, fuelled by the media. The nature of the death can further amplify this. Given that those such as Kluckhohn[30] have suggested that culture can be defined as a response to 'problems', studying the response to what could be considered the ultimate problem – an 'untimely death' – may be especially instructive in learning about the culture of a particular people. Japan is a country which is ripe in opportunities for studying untimely deaths. Historically there were deaths due to *seppuku* (*hara-kiri*), where, although those conducting the practice may have been prepared, their family may not have had such time. Similarly there were the *tokubetsu-kōgekitai*, better known as *kamikaze* pilots. More recently, Japan has become the focus of international media attention for Aokigahara, a forest at the foot of Mount Fuji where an average of 30 bodies of those who have committed suicide are found each year,[31] internet suicide groups,[32] and suicides on the railways. Japan was also the focus of the world when 12 people were killed and many hundreds were affected by the sarin gas attack on the Tōkyō underground in 1995. Furthermore, as highlighted in Chapter 2, there are many deaths due to 'natural disasters'. But the case of JL123 reflects the fragility of the lives of those living in a modern, developed society that is dependent on technology and mass transportation.

Most Japanese seem to have a greater acceptance of death as being a part of life than many in Britain, for example. Perhaps this is due to the country frequently being struck by 'natural disasters' and so they also know that death could happen at any time. Rather than characterizing the Japanese attitude as being fatalistic, I would describe it as being realistic. But this realism does not mean that they are unresponsive to death. The rituals and memorialization processes that this chapter highlights are a means to address and express their feelings.

Trying to understand death and religion tends to revolve around questions such as 'Who am I?' and 'Where did I come from?' I cannot separate the physical and non-physical parts of me. I am aware that many of my non-physical characteristics are similar to those of my parents, and that some have even developed in the years since living away from them. So, this non-physical side of my existence owes as much to them as the physical. In turn, I owe much to my ancestors. I find it amazing that I contain DNA that has been passed down thanks to ancestors surviving wars, managing to find food and perhaps even escaping from wild animals, and that if this had not happened and various combinations of partnerships had not happened then I would not exist.[f] But ancestor worship is not important to me. I am largely comfortable with the idea of 'you come from nothing, you're going back to nothing', as Monty Python's *Always Look on the Bright Side of Life* puts it. That is to say, that at death, everything stops. However, when I consider the bigger picture, in terms of how the Earth, the galaxy and the universe were created, and what was there before their creation, I have much more trouble coming to a satisfactory conclusion. Where did all these atoms, molecules and elements originally come from? What is on the other side of the edge of the universe? Science has no definitive answers. So perhaps there is some kind of entity that created everything. In which case I cannot ignore the possibility that there may be

Eiji Nakagawa 中川英二 *(38)* · *Hidenori Nakagawa* 中川秀則 *(41)* · *Takeshi Nakai* 中井敬 *(28)*

some other existence after our physical death on Earth and, by extension, it is then conceivable in that existence that we might 'meet' others who previously existed on Earth. Consequently, I am mindful of the possibility of one day meeting those who were on JL123.

Although I probably fall into the category of having 'no religion' and not having a 'belief' in the sense of many of the world's main religions, my life is not devoid of religion, I just have no need to follow an official religion. Of course the issue is, as discussed above in relation to Japan, what is meant by religion. If going to a place where one engages in forms of ritual singing, prayer and even worship is what is meant by religion, then support of a particular football team, for example, for many must border being on a religion.[g] The state is usually only concerned with officially recognized religions. Yet, for many people, there are many other aspects of their lives which could, and in many societies would, be considered as religious. In some societies, some practices may get labelled under the banner of 'superstition' by others. But where is the dividing line between 'superstition', 'common practice', 'religious practice' and 'religion'? Japan and the aviation world also have numerous 'superstitions'. In Japan the numbers four and nine are unlucky. Four is unlucky as one reading for the *kanji* is *shi* which is also the word for death, albeit written with another *kanji*. Nine is unlucky as a reading for the *kanji* is *ku* which can also mean suffering when written with another *kanji*. Japan has also adopted the number 13 from many Western cultures as being unlucky. If one looks at the seating plan of JL123 (Figure 3.5) you can see that there is no row 9, although there are rows 4 and 13.[h] Of course, this means that what is written as row 10 is actually the ninth row. Most people are not particularly unlucky on Friday 13th, a date considered unlucky in many Western cultures, but despite this, the next Friday 13th may lead to unease, and anything unlucky that does occur may be remembered over and above the number of 'ordinary' Friday 13th they had experienced before. In relation to JL123, 12 August 1985 was *Senshō* according to the *Rokuyō* calendar of six auspicious days.[i] This has the meaning of being lucky in the morning, but not in the afternoon. There are also those who believe that clues to JL123's fate can be found in astrology charts.[33] Furthermore, while a proposal for the second mass cremation on Friday 13 December was rejected, Kawamura points out that this was not to with any issues with Friday 13th itself (see also Chapter 5).[34] Other issues relating to numbers will be returned to in Chapter 8.

JL123 memorialization in 1985

As explained above, for many Japanese there is a belief that the individual does not die at the time of the physical death and that certain rituals should be performed to aid the individual's transition to becoming a Buddha. These rituals and beliefs drove some of the events to take place in 1985 and also the way in which the remains were considered.

The term for a dead body in Japanese is *itai*, made up of two *kanji* – *i* (leave behind, bequeath) and *tai* (body). The *i* is the same *i* that forms the term *izoku* (literally 'family left behind'), *ihin* (belongings left behind) and *isho* (a written document left behind, or a will). The nuance of this *kanji i* and the word *itai* appears different to some other cultures. Miyake argues that at the heart of the difference is the Confucian idea that the body is a gift from one's parents and ancestors,[j] and that the duality of spirit and body, whereby the spirit is 'imparted to the living by God' and the body is 'of little consequence', does not apply.[35] But I fear that sometimes this view is too simplistic and exaggerates the situation. It also does not apply in all cases. Iizuka tells that one foreign *izoku* of JL123 suggested that they did not need their son's body returned and what was important that his spirit was with God.[36] But having read this *izoku*'s diary of their time in Fujioka, spoken to them and also to the JAL employee looking after them, I would suggest that this is not an accurate reflection of what this *izoku* felt and I doubt such a thing was ever said. Similarly, if it were really the case that the Japanese wanted all remains to be identified, as Miyake suggests, then surely the identification process should have been conducted in a way conducive to doing so (see Chapter 5). Whilst Yoshioka points out that some *izoku* did want all their loved ones remains returned so that they could 'eat their food' or 'cross the Sanzu River',[k, 37] for most this was clearly not an issue. Yet, although Iizuka and Miyake may have exaggerated the situation, there are clearly differences between some Japanese and some Christians.

After the identification of JL123's victims, there was a cremation and funeral. Although cremations have become the norm in Japan, this is a relatively recent development. It was not until the past 100 years when there were increasing concerns about the lack of space for burials, where the coffin was inserted vertically rather than horizontally, which was the norm in the growing urban areas that cremation were adopted as a pragmatic solution.[38] However, cremations in Japan are carried out at a lower temperature than in some cultures, allowing families to take a selection of bones away for inclusion in the ancestral grave. That people take part of the deceased with them is a further indication of how, despite the Shintō influenced ideas of avoiding death, there are examples of Japanese practices where death is embraced, albeit within certain limits, more than in many other societies. Indeed, a German *izoku* was probably shocked to be greeted and handed an urn containing the cremated remains of their loved one, given that in Germany, particularly in the 1980s, handling the urn was almost unheard of.[39]

In the case of JL123, 87 per cent of the 518 victims which were identified had their remains cremated in Fujioka, Takasaki, Tomioka, Maebashi, Isezaki or Kiryū in Gunma prefecture, and the towns of Okabe and Misato in Saitama prefecture near the border with Gunma (see Figure 6.2). Others were cremated in their home prefecture, after the bodies were transported back there. In September, discussions about a mass cremation began.[40] Although JAL proposed 2 October, this was not possible due to the opening of a new expressway that day and the logistical problems it would cause for the Gunma Police,[41] and so it was held on 5 October.[42] In total 893 remains were gathered into 29 coffins to be cremated in five locations in Gunma.[43] The ceremony was not without its problems, however, as some *izoku* complained after Ueno-mura mayor Kurosawa thanked JAL for

Minako Nakajima 中島美菜子 (0) · Hiroshi Nakajima 中島博 (36) · Makoto Nakajima 中島誠 (38)

their help in the relation to the event.[44] With the identification process coming to end and there being little, if any possibility of further victims being identified, a date was needed for a second mass cremation. Although the original plan was for the second cremation to be in the middle of the month,[45] many *izoku* wanted it to be after 18 December. The reason for this was that there would be a JAL shareholders meeting on that date when Susume Yamaji would take over from Yasumoto Takagi. That Takagi had not been at the first ceremony was an issue, and many *izoku* wanted to ensure that the new president did not forget about JL123.[46] As it was 20 December was decided upon and Yamaji was at the front of the building to bow to all *izoku* as they entered the building.[47] As well as the two mass cremations, there were also two memorial events – one in Ōsaka on 22 October and one in Tōkyō on 24 October – both organized by JAL.

JL123 reveals a number of issues relating to symbolism and death in Japan. First, there is the way in which the coffins were displayed within the makeshift morgue and the morgues themselves (look back to Figure 5.7). The coffins were draped in a white sheets. White is the colour generally associated with death in Japan, although it is still commonplace for people to wear black for a funeral and a bride to wear a white wedding dress. On top of the coffins, as well as the identification sheet and some belongings *(ihin)* that were known to be those of the victim, was a bunch of white flowers tied with a white bow. When one considers the heat in the morgues and the number of days, and in some cases, weeks that

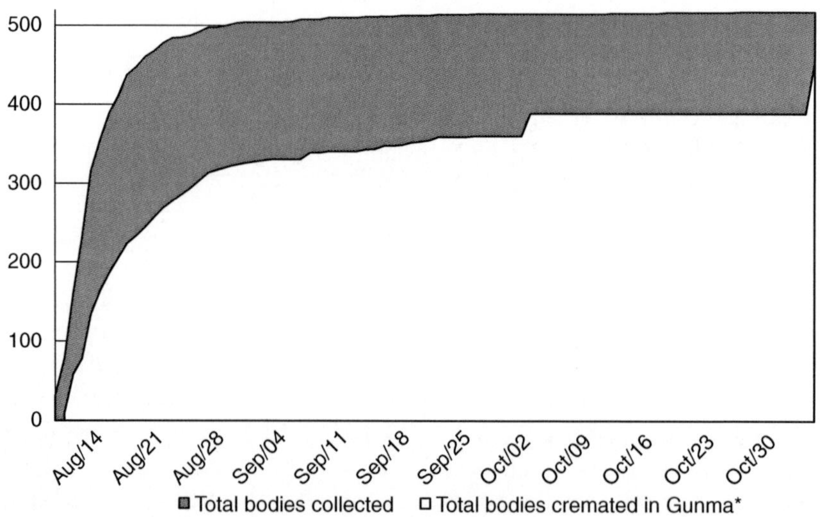

Figure 6.2 Collection and cremation

Figure by the author, based on data in Gunma-ken Sōmubu, 1986:48–53.

Note: * Also includes data for 20 cremations that took place in Okabe and Misato in Saitama prefecture.

Figure 6.3 Altars in the home

Photographs courtesy of Peter Mathews.

these coffins were kept there, it must have been a challenge to ensure that there was a sufficient supply of fresh flowers. As well as the flowers placed directly on the coffins, there were also flowers on display around the morgue itself. Generally these were presented in a large circular wreath, known as *hanawa*, with the name of the person who had sent them written below.

Another feature of symbolism and death in Japan can be seen in the way photographs are used. Most constructed a temporary altar in their home (see Figure 6.3a), which would be replaced by a more permanent altar, known as a Butsudan (see Figure 6.3b).[1] Food and drink are usually placed in front of the Bustudan. These offerings are another example of the blurring of the distinction between the living, the physically dead and the recently dead, as discussed above. Generally, a black and white picture showing the face and upper body of the deceased in formal wear is displayed. In an age of colour images, to ensure there is a black and white image suggests that a particular conscious effort is being taken. It would seem that the use of black and white helps to signify the passing of the individual from the colourful, living world, as black and white pictures, at least since the development of colour film, are associated with the past. Another conscious action that tends to follow a death is revealed by Iwayumi Suzuki, who points out that photographs containing both the living and the deceased are generally not on show in Japan. Somebody has to take responsibility for ensuring that no such pictures are on display.[48] For many this will be relatively straightforward, as Suzuki also notes that the average number of pictures on display is as low as 3.4 per household, and nearly half of photographs on display are of dead relatives.

Photographs can also serve another purpose. I have seen photographs of *izoku* taking the prints of JL123 victims to a baseball game. A strong link developed between JL123 and the Hanshin Tigers baseball team as the president of the team,

Yukiko Nakamura 中村悠紀子 *(38)* · *Toshitaka Nakanishi* 中西稔剛 *(38)*

Hajimu Nakano, was on board JL123 and many of the victims were Hanshin Tigers fans. Consequently, many people around Japan began supporting the Tigers, keen for them to win the championship that year. Such support was also seen in the wake of the 1995 Great Hanshin Earthquake.[m] Taking photographs of victims of the JL123 crash to a baseball game was therefore a natural extension of the support being given nationwide to the Hanshin Tigers. It could even have been an attempt to ensure the victims did not miss out on the Tigers winning their first Japan Series and first Central League championship since 1964.

In relation to death in Japan, one of the most significant times in the year is 'Bon' or 'Obon', a three-day period which can be best translated as 'Festival of the dead'. The exact timing of Obon varies from region to region in Japan. In the big cities of Kantō and in Tōhoku it tends to be focused around 15 July, while in others areas of Japan it tends to be focused around 15 August, but there are yet more variations. Being a three-day festival, and with there being a reluctance or inability to take long holidays in Japan, it is inevitable that 12 August becomes one of the busiest days for travel in Japan. There are particular rituals associated with Obon, including the *Bon Odori* (the bon dance), *haka mairi* (the visiting and cleaning of the family grave) and *tōrōnagashi*, when lanterns are lit and floated down a river (see below). Obon itself is not generally a sad event, but a celebration of life and the gratefulness to ancestors.[49]

Many of the death rituals discussed could largely be seen as being positive actions – trying to help a loved one become a Buddha and ultimately an ancestor. However, there are aspects of the rituals related to death which could be better described as being defensive for those left behind. The concern, whether fully believed or not, relates to whether the spirit of the deceased will be able to progress to the next world and that if they do not, they may become a vengeful, evil spirit or ghost (*yūrei*) and haunt the living.[50] The use of rituals and memorials is one aspect of this and they are particularly significant when the individual has suffered an untimely death. The dividing line between it being a positive action or defensive action would be hard to define, but we need to be mindful that the rituals are not done totally selflessly. It is estimated that the majority of memorials which can be seen in the street in Japan are for those who have suffered an untimely death,[51] indicating the level of concern that may exist. The memorials in Ueno-mura can be seen as a further example of the memorials which could be designed, not merely to remember those who have died, but as an attempt to placate the spirit of those who have died.

Japanese culture has many ghost stories,[52] and Japan's horror film genre is internationally renowned. In 2010 Atsugi city government even had to buy a disused hospital building because of the stories about it being haunted, and it was torn down and the area turned into a park.[53] Despite the lack of evidence in the existence of ghosts, many seemingly perform rituals or behave in a particular way just in case. Although there are internet discussion boards which include questions about whether or not anyone has seen a ghost at Osutaka-no-One, there are no reports of any observations. Perhaps this is due to the fact that the JL123 *izoku* have performed the rituals so well. When I went to sleep by myself in one of the *yamagoya* (mountain hut) at the crash site, many suggested that I may see a ghost.

Although probably said in jest, the comments do indicate a certain uneasiness which surrounds sites of death. However, there are ghost-related stories related to JL123, for example, an odd sighting near the crash site was suggested by some as being a ghost.[54] But, as Iizuka points out,[55] it is most likely that this was a group from Ritsumeikan University who had tried to find the site after seeing the news whilst staying in the area.[n] The owner of a *rāmen* (noodle) shop next to one of the buildings which was used as a morgue reported that he lost business due to rumours about ghosts being in the area.[56] The building used as a morgue, the Shimin Kaikan, was pulled down by Fujioka city government and replaced by a new building with only a memorial stone left on the former site.

The Osutaka Pilgrimage

Many *izoku* go to Ueno-mura in August each year to attend commemorative events held there. I refer to this as the 'Osutaka Pilgrimage' as it reflects many aspects of other pilgrimages in Japan,[o] but it is necessary to clarify what is meant by the term 'pilgrimage' and the particular features of many Japanese pilgrimages.

Pilgrimage is generally understood as being a journey to a location for the purpose of a religious experience of some kind and features in most major religions. It involves a break from one's normal way of life and the journey itself may be arduous. Kitagawa points out that

> In every religious tradition, the pilgrimage combines, more than other religious acts do, diverse and often contradictory features, which are both spiritual and mundane. Travelling a long distance, visiting holy mountains or shrines, involves physical hardship and endurance, but it also has pleasurable aspects, such as sightseeing and meeting new friends.[57]

In many cultures a pilgrimage is made to a single location, whether it be Mecca, Jerusalem, Canterbury, Lourdes or Santiago de Compostela. But, as many, such as Reader,[58] Walter,[59] Moore,[60] Notoji[61] and King,[62] have pointed out, pilgrimage need not be to a religious location, although the experience once there may be comparable to a religious one (keeping in mind the discussion elsewhere in this chapter about the problems of defining 'religion' and 'religious'). Pilgrimage is also about self-discovery and 'spiritual' development.[63] Reader suggests that,

> the idea of a journey out of the normal parameters of life, the entry into a different, other, world, the search for something new, the multiple motives of participants, ranging from homage and veneration to the simple impulses of curiosity are extremely common, if not always universal, denominators of pilgrimage.[64]

So, visits to war graveyards,[65] Disneyland,[66] Graceland[67] or even a football ground may be construed as being a pilgrimage. As West Sellars and Walter suggest, 'pilgrimage is a matter of popular culture – nearer folk religion than official religion.'[68] Although the term 'folk religion' can be problematic as it suggests 'primitive', given what has already been discussed here about the way in which religion in Japan is practised, one can understand why pilgrimage may be attractive to many Japanese. Furthermore, Reader notes that there is often a 'close relationship between pilgrimage, heroic figures and death, *especially sudden and violent death*' (emphasis added).[69] This point is significant in relation to JL123 and the Osutaka Pilgrimage.

There are many different types of pilgrimage in Japan and different words (including *mairi*, *mōde*, *junrei*, *okagemairi*, *henro*, *sankei* and *kodō)* used to describe them.[70] There are several hundred pilgrimages in Japan and most are small and local, but some cover much larger geographical regions.[71] As was discussed in Chapter 2, historically this was one of the only reasons for being able to get permission to travel at all, and so pilgrimage and travel, and, by extension, tourism have been entwined. Indeed, Turner and Turner noted that, even in Christian culture 'a tourist is half a pilgrim, if a pilgrim is half a tourist'.[72] This has meant that although, particularly in the past, there may have been an expectation that it would involve some form of 'religion' or 'self-discovery', pilgrimage can also contain ludic elements. Another important feature of many Japanese pilgrims, particularly historically, is that they were easily identifiable as being such. They were making a break with their normal life and this was signalled through the clothes that they wore (see below) – so that they were 'dressed for and ready for death at any time: symbolically s/he is dead to the mundane world'.[73] Today it is rare for pilgrims to dress like this, although some will carry 'some token of white' on them,[74] whilst others may carry the bones, ashes, or *ihai* (mortuary tablet with the posthumous name on it) of a loved one who has died.[75]

In comparison with many other cultures' pilgrimages, a significant difference is that most Japanese pilgrimages, since the first pilgrimages of the ninth century,[76] involve visits to more than one location. The most famous pilgrimages are the Shikoku Pilgrimage and the Saikoku Pilgrimage. The former involves visiting 88 temples on the island of Shikoku,[p] and is promoted 'as a symbol of Japanese tradition and cultural heritage'[77] and a 'Japanese spiritual homeland'.[78] The latter involves visits to 32 temples around the Kansai region.[79] Even modern day, less obviously religious pilgrimages will often have multiple points – a visit to Tokyo Disneyland (which has multiple areas and rides) being combined with visits to other sites in Tōkyō, including particular restaurants; and trips to an *onsen* resort will often feature visits to more than one *onsen*, for example.

Although many modern-day Japanese pilgrimages do not have an overly arduous element to the journey – there are even some who take the Shikoku pilgrimage by helicopter[80] or merely 'visit a temple' by going to an exhibition in a department store where soil, for example, from the temples can be 'visited' – the Osutaka Pilgrimage still involves a difficult journey. Chapter 4 has already highlighted many of the issues relating to accessing Ueno-mura and Osutaka-no-One. In terms of getting there today, although the roads have improved significantly since 1985, it is still a difficult drive and there is very limited public transport. For those coming from Kansai, arrival in Tōkyō is about the three-quarters point in terms of distance, but they are barely at the halfway point in terms of time, depending on the exact route taken. In Ueno-mura today there are a number of options for accommodation, although one of the two largest was not established until 1995 and the other was rebuilt in 2001. In total Ueno-mura can accommodate around 500 people per night.[81] Before then, staying in Ueno-mura was a much greater challenge for the *izoku* and tended to restrict most to making a day trip there. With the final ceremony on 12 August not finishing until about 19:00, and with the time needed to access Osutaka-no-One (which used to be much greater, see below), it made it a very long, difficult day. That everything had to be done on one day also restricted what events could be conducted.

As noted above, one of the aims of the 8/12 Renrakukai is to 'comfort the spirits' of the victims, so it is no surprise that the organization has been actively involved in the remembrance events in Ueno-mura. There are now three main events and it is easiest to understand these in the chronological order in which they take place on 11 and 12 August each year.

Since 1995, on 11 August, the *izoku* hold a service known as *tōrōnagashi* (see Figure 6.4), which is a feature of many Obon festivities and is also included as part of the memorial events for the atomic bombing of Hiroshima. Although a dictionary definition of *tōrōnagashi* is 'lantern festival', this does not really capture the nuance of the event or provide an understanding what it involves. From around 16:00, *izoku* begin gathering on the river bank in central Ueno-mura near the village hall (*yakuba*), which is where the accident response activities were co-ordinated from in 1985.[q] Lanterns are laid out along the flat, stony riverbank. Some of the *izoku,* staff from JAL and other volunteers then prepare an area so that everyone will be able to place their lantern in the river. Another area is prepared to help capture the lanterns further down the river, for, unlike the case of Hiroshima, for example, the lanterns are taken out of the water and not left to continue downstream. After this has been prepared, a few lanterns are placed in the water to ensure that it works properly.

From around 16:30 people begin to write messages on the lanterns. Most of the lanterns are only about 20 cm tall and have two blank panels on them for people to write messages, and four panels with printed drawings on. There is also a number of larger lanterns which have messages already written on them; one of these is designed in the shape of Irei-no-Sono (see below), another lists other disasters which are being remembered at the event (see below). As the messages are written on the lanterns, volunteers of the Takasaki Accordion Circle provide musical accompaniment. At 18:20 candles in the approximately 300 lanterns are lit. At this point it is starting to get dark in Ueno-mura so the light glowing from inside the lanterns can easily be seen. People pick up their lantern and begin to form a queue. A number of speeches are made at this point and a minute's silence is held, followed by the singing of songs, including *Sen-no-Kaze ni Natte* (see Chapter 1). After the songs, the process of putting the lanterns into the river begins. When all the lanterns have taken their journey along the river, some further speeches are made, songs are sung and bubbles are blown. By this time it is around 19:00 and

Tadao Nishikawa 西川忠男 *(42)* · *Hiroyuki Nishikawa* 西川博之 *(28)* · *Shōji Nishimura* 西村庄二 *(52)*

Figure 6.4 Tōrōnagashi

Photographs by the author. a) Setting up along the river bank in front of Ueno-mura village hall; b) the media focus on *izoku* and others write messages on lanterns; c) *izoku* and others queue up to put lanterns in the river; d) one of the large lanterns to the spirits of the 520 victims; e) a large lantern listing some of the other tragedies being remembered; f) a large lantern in the shape of the Irei-no-Sono memorial; g) a large lantern calling for air safety on one panel and with Disney characters, as a reminder of the children who died on JL123, on another; h) lanterns floating down Kanna-gawa; i) the media in the river taking pictures; j) speeches, songs, accompanied by the Takasaki Accordion Circle, and the blowing of bubbles.

pitch black, with much of the light being provided by lights used by the media who cover the event each year (see Chapter 7). The *izoku* then go to dine and stay in the local accommodation, and many also take advantage of the *onsen* there.

Much of what takes place during *tōrōnagashi* is very symbolic. The idea of *tōrōnagashi* is that the spirit of the deceased is able to be contacted through the lanterns and that they will read the message. I doubt that there are many who really believe that this happens, but in the case of the JL123 *tōrōnagashi*, where the lanterns are in the water for only a few metres which takes barely a minute to cover, and are always in sight of the people congregated there, I suspect this level of belief is even lower. But, as has been discussed above, much of what happens in relation to remembrance is about conducting the appropriate rituals rather than the actions being a reflection of belief in what is being done. The act of demonstrating a commitment through partaking in the rituals is perhaps of greater significance, just as the displaying of an amulet (see below). The actions are a symbolic act that reaffirms the memory of the deceased to the living, and, particularly in the case of JL123, reaffirms the living person's identity with the crash and perhaps also the 8/12 Renrakukai. The symbolism of the *tōrōnagashi* is probably enhanced by the fact that the river, Kanna-gawa, is fed by a number of streams from within Ueno-mura, including Sugeno-sawa, where much of the wreckage of JL123 came to rest. That the river's name literally means 'deity flowing river' may add extra poignancy. Indeed, this relates to another aspect of 'religious belief' in Japan – that spirits are thought to reside in the mountains, returning to their village during Obon,[82] and as such mountains have always been important religious places in Japan,[83] as well as places associated with death.[84] That JL123 crashed in mountains around the time of Obon, therefore, may add significance for many Japanese.

On 12 August memorial services are held at Osutaka-no-One and Irei-no-Sono. Over the years, the road to Osutaka-no-One has been improved. The cost of this was around ¥1.2bn, with around ¥1bn coming from JAL.[85] In 2005 a new dam was built in Ueno-mura which required a new road, with tunnels, to be constructed. This has further improved the route between the centre of the village and the crash site. Many believe that this was the real objective of building the dam,[86] and the fact that the mayor of Ueno-mura was successful in getting the project realized is a case of when the localized nature of Japanese politics and administration can have positive aspects. Despite these changes, it is still a difficult drive and the 20 km journey can take 40 minutes. Referring back to Chapter 4, this journey could have been very different if village and prefectural boundaries were not taken into consideration and a route had been built from Kawakami-mura, although this village itself may be harder to access. However, it is unlikely that many would have desired the other memorial, Irei-no-Sono, to have been built in Kawakami-mura given that it had no association with the crash. The problems of the route from central Ueno-mura to Osutaka-no-One were highlighted in 2008 when it was closed due a landslide.[r] The subsequent repairs and improvements cost ¥50m.[87] Alterations have also been made to the car park at the end of the road. A new car park was opened in 2006 which halved

Tatsuo Nishiyama 西山竜夫 (33) · Toshie Nishiyama 西山俊江 (and foetus) (26)

the time of the uphill walk to the site.[88] Despite this, the final walk is still 800 m and involves a climb of 182 m in that distance. Although not difficult for the average adult, it is physically challenging for many young and elderly visitors. The challenge for older *izoku* has been a particular concern, and as a result, volunteers are available to provide a massage as they return to the car park after visiting Osutaka-no-One. One of those concerned with the challenges facing the older *izoku* was the Governor of Gunma,[89] who implemented the installation of new barriers and steps in 2009 at a cost of ¥20m.[90] That this action was needed was proven in their first year as heavy rain fell on the night of 11 August, and without the new steps it would have been almost impossible to reach the focal point of the Osutaka-no-One by the direct route, although it may have still been possible using the longer, steeper, route via Sugeno-sawa.

The main focal point at Osutaka-no-One is located where the heliport was made following the crash (see Figure 5.2). Around this area there are memorial stones, a bell, places for people to put the *tanzaku* (paper with messages) provided and a *saidan* (altar) hut for people to leave artefacts (see Figure 6.5i). Many of the *tanzaku* have messages on them calling for improved aviation safety or a reinvestigation into the cause of the crash, the hope that the victims rest in peace, recollections of what the individual was doing that day, or messages to individual victims (with a number being to Kyū Sakamoto). There is also a statue of Kannon, the deity of mercy (see Figure 6.5h), who is thought to be able to 'confer this-worldly benefits and to intercede in the form of miraculous interventions that save people from danger or even death'.[91] The hut is perhaps reminiscent to the way in which pilgrims may leave *tsue* (a staff, see below) and other items during the Shikoku Pilgrimage.[92] Although these are sometimes cleared away, only to be replaced by new ones by other pilgrims,[93] I have not noticed any significant clearing out of items in the hut at Osutaka-no-One. It contains photographs, mementos, *senbazuru*[s] and notes from visitors – particularly non-*izoku*, including those who appear to have come as part of a group from a company, organization or union.

The main memorial stone is Shōkon-no-Hi ('stone monument to the risen souls') (see Figure 6.5b), another records the date and time of the crash (see Figure 6.5f), and a further memorial stone has the names of the 520 victims on it (see Figure 6.5g). The number of memorials has increased over the years, with the bell added in 1993, supposedly used to notify the deceased that their loved ones have come to visit. Although many do ring the bell, there is no consistency in its usage; some will just give it one ring, some will ring it several times, and many others do not use it at all. In 1999 the IREI Air Safety planted a 'time capsule' containing *ihin* and other items at Osutaka-no-One.[94] Subsequently, another memorial with nine busts and a small bell on it was added in 2001 (see Figure 6.5e). The practice of having statues with a likeness of someone who has died goes back to the Meiji period,[95] but the fact that only nine victims are depicted speaks to the lack of common response to the JL123 crash. The site is adorned with flowers, particularly around 11 and 12 August. Non-*izoku* visitors to Osutaka-no-One will bring flowers to this spot as they have no other focal point to place them.[96] However, the flowers have become increasingly popular with

Figure 6.5 Memorials at Osutaka-no-One

Photographs by the author. a) a memorial at the start of the climb up Osutaka-no-One; b) the focal point of Osutaka-no-One, in the centre is Shōkon-no-Hi, in the foreground is the bell and places for messages to be left; c) a collection of stones for the bodhisattva Jizō; d) a memorial near Shōkon-no-Hi established by *Kōkū Anzen Kokusai* Rarī – in the foreground is where a time capsule has been placed and in the background (shown larger in picture e) a bell and faces of nine people; f) a memorial stone commemorating the exact time of the crash as well as details about the number of passengers, crew, victims and survivors; g) a memorial stone with names of the 520 victims on it; h) a statue to Kannon, the deity of mercy; i) a hut with various items which have been left by visitors to Osutaka-no-One.

the local wildlife, such as deer, and much time is needed to make sure that the whole site looks pleasant.[97] In addition to cut flowers, other plants and trees have been planted, including *sakura* trees,[t] although they rarely blossom because of the cooler climate at this altitude.[98] Some will also light *kasen* (incense sticks) near *Shōkon-no-Hi*, although care needs to be taken with the fire risk (see Chapter 4), and people are requested not to use them at individual *bohyō* (grave marker).[u]

Many of the *izoku* gather in front of Shōkon-no-Hi at 10:30 on 12 August. As with the *tōrōnagashi*, the service includes speeches, songs, musical accompaniment from Takasaki Accordion Circle, and the blowing of bubbles (see Figure 6.6). The 8/12 Renrakukai appears to be careful to get a variety of its members involved – for example, different people make the speeches at the *tōrōnagashi* and the Osutaka-no-One service. One of the songs sung at the service, also used at the *tōrōnagashi*, is particularly poignant at this service. It is called *Ā Osutaka-yama* and was written by one of the *izoku*. The blowing of bubbles at both services is something that has apparently become more widespread at remembrance events in Japan.[99] In the case of JL123, that there were so many child victims is another reason why the bubbles are used, a point that was emphasized during one of the speeches in 2007 which also mentioned the need for a 'baton-touch', like a relay, to the next generation to ensure that the remembrance events and activities of the 8/12 Renrakukai continue.

As well as the collective memorial service, families will also spend time individually at the place on Osutaka-no-One where their loved one died. Some families do this before the memorial service, others after, and some do not even attend the service itself. Those that go to the individual spot may spend an hour or more there, cleaning the area, having food, reminiscing and talking to other family members and updating their loved one on the family's lives since their last visit. They may also be interviewed by the media (see Chapter 7). The musicians of the Takasaki Accordion Circle also go around the site and play requests by the *izoku* for their loved one(s).

The site itself has changed over the years. The most significant aspect is that many trees have grown. Although there was some discussion about not allowing trees to grow back, in the end it was decided to allow most to grow back and others to be planted. A comparable discussion took place about whether to maintain the whole of Hiroshima in its 6 August 1945 state or allow the city to be rebuilt and only maintain a few parts as a reminder.[100] The trees at Osutaka-no-One are not only a reminder of the continuation of life but also provide valuable shade from the heat of the summer sun. The site is also relatively pleasant as the shrill of cicadas seems less obvious than down in the cities, where they tend to be concentrated on the few trees which remain in the urban landscape. Osutaka-no-One is much like a wooded park on a hill, albeit punctuated with memorials, so one cannot help that the visual impact of a bare ridge would be greater. That the trees have grown up also makes it harder to see the site as it is hard to view much of it from any one place; but it is important to remember that most visitors to Osutaka-no-One are *izoku* and so the 'visual impact' is perhaps a minor consideration compared with the comfort of the more regular visitors. However, in 2010, the

Figure 6.6 Memorial services at Osutaka-no-One

Photographs by the author. a) the *yamagoya* on Osutaka-no-One provide a place for *izoku* and others to rest; b) new handrails and improved steps installed in 2009 have made the climb easier; c) a Buddhist priest pays his respects during the early morning mist on 12 August; d) two *izoku* pay their respects at Shōkon-no-Hi as the media watch on; e) a child is among those making a speech during the morning memorial service at Osutaka-no-One; f) the Takasaki Accordion Circle go around the site to provide musical requests for *izoku* by particular *bohyō*; g) the media frenzy at Osutaka-no-One around the area of Shōkon-no-Hi; h) the media following *izoku* to *bohyō* in Sugeno-sawa; i) balloons and small windmills are a new addition to the memorials in 2010 – in the background is the U-shaped cutting on the far ridge; j) the JAL president weeps as he gives an interview after paying his respects at Shōkon-no-Hi in 2010; k) the media crowd around Transport Minister Maehara as he gives an interview after he has paid his respects at Shōkon-no-Hi and before being shown around most of the crash site.

Miyajima's cut down some of the trees near their son's *bohyō* so that he could see the beautiful view across the nearby mountains.[101]

Walking around the site there are many reminders of what happened in 1985. From Shōkon-no-Hi and looking across to the next mountain, it is still possible to see the U-shaped cutting created by the right wing of JL123 (see Chapter 3). Although the trees have grown up, due to the depth at which the ground was cut, the trees have not reached full height and so the gap remains visible (see Figure 6.7). Near the top of Osutaka-no-One is a rock, referred to *X-iwa* (*batsu-iwa*, Rock X) against which the front part of the plane impacted during the crash. It is still possible to find fragments of the plane, particularly after the soil has been washed away by heavy rain.

However, the main reminder of the enormity of events on Osutaka-no-One is provided in the number and design of *bohyō* and what is left by them around the site. As mentioned in Chapter 5, marker posts were placed at the site in the spots where remains were found. A metal tag on these posts indicates whose remains were found at that spot. Subsequently, *bohyō* were placed by *izoku*, or by JAL staff on their behalf, next to the metal posts. In many cases the metal post has

Figure 6.7 Scars at Osutaka-no-One

Photographs by the author. a) the U-shaped cutting created by the right wing of JL123 on another ridge before it hit Osutaka-no-One; b) the 'Rock X' which the front part of the plane hit during the crash; c) a tree scarred by the crash; d) a new tree growing up through the centre of a larger, hollow, tree damaged by the JL123 crash; e) a fragment of JL123 at the site.

been removed, but a large number remain. There are also cases where there are no *bohyō*, and at least one case where no post or *bohyō* exists. The design of *bohyō* varies considerably; many having been updated on several occasions; others are hard to read having seemingly not been replaced since they were first placed. Let us now consider these variations in more detail.

Figure 6.8 shows how diverse the designs of the *bohyō* are. Although perhaps hard to see the image in detail,[v] the two-page spread is intended to give an impression of both the diversity of designs as well as to highlight some of the common features. While most of the *bohyō* are wooden, there are some stone and marble constructions. The use of stone would suggest that the *izoku* wants to 'pass down a memory forever'.[102] Some of the *bohyō* contain photographs of the victim. Given what was written earlier about photographs, it is perhaps surprising that some of these are colour. Most *bohyō* are relatively plain and have the name of the victim(s) written on them, but there are examples where there is just a message. Although one may expect to see the posthumous name (*kaimyō* or *hōmyō*) used on *bohyō*, my survey has found that some 87 per cent of the *bohyō* have the victim's secular name (*zokumyō*) written down. There are examples of *bohyō* for a whole family without the name of each individual being given, but there are many more examples where each individual is named on the same *bohyō* or each individual member of the family has their own *bohyō*. There are also numerous crucifixes – for both Japanese and non-Japanese victims – while other memorials display a more overtly Buddhist influence. The *bohyō* for the captain, co-pilot and flight engineer are part of one construction, although there are three individual posts.[w] There is one case where one victim has two *bohyō* in different locations at the site. One, placed in 2009, is located where the remains were found, the other is located near other *bohyō* as the victim's son was apparently concerned about his father being by himself.[103] We should remember that in addition to the *bohyō* at Osutaka-no-One, the victim will also be listed at the family grave in their home town, and as such the translation 'grave marker' for *bohyō* is misleading.

Items are often left to the *bohyō*, for example I have observed flowers, drinks and food, Jizō statues,[x] photographs, toys, a game of *Go* (an ancient board game), a golf ball, Disney characters carved in stone, and Hanshin Tigers baseball caps and flags. I have even come across an umbrella being put over the *bohyō* – whether this is to protect the *bohyō* itself, the other items left there or even the individual is unclear. The use of toys and items for children highlights the fact that the victim ceases to age in the minds of the *izoku* and is effectively frozen in time. But whilst *izoku* go there to remember the past, there may also be 'nostalgic yearnings are for what might have been', as Reader has noted it in relation to other young people who had an 'untimely death'.[104] In the case of photographs, some are family pictures which include the victim, whilst others appear to be for the victim to see how the family continues. That there are some of the former indicates that there are times when the taboo of having pictures on display which contain 'the living' and 'the dead' together breaks down. Next to many *bohyō* are wooden *sotoba* posts, which have a Buddhist inscription written on them. New ones are brought on each visit, with older ones pushed to the back to decay naturally. This sight is

Figure 6.8 Bohyō at Osutaka-no-One

Photographs by the author. 1) *bohyō* wrapped with a Hanshin Tigers towel; 2) *bohyō* with a message, 'Father please rest in peace' on the tablet by the side; 3) the stone has the first name of the victim a the message, 'A kind father'; 4) 'Father rest in peace'; 5) a stone tablet gives information about the victim and even the date of the crash; 6) *bohyō* for the cockpit crew; 7) a modest plastic name plate; 8) a foreigner's *bohyō*; 9) Hisashi Ōshima's (a.k.a Kyū Sakamoto) *bohyō*; 10) a drink is left by the stone memorial, surrounded by *sotoba*; 11) 'Thank you for your love'; 12) *bohyō* for a whole family; 13) 'Our father sleeps here'; 14) 'Father, we'll come again'; 15) *bohyō* surrounded by Buddhist memorials; 16) Hajimu Nakano – 'The place he rose to the heavens'; 17) *bohyō* with posthumous name on the front and original name on the side; 18) a picture of the victim left by a stone memorial; 19) *bohyō* with more elaborate calligraphy on it; 20) a collection of toys left for some young victims; 21) *bohyō* covered by an umbrella; 22) *bohyō* with metal calligraphic name plate on it; 23) a less regularly shaped *bohyō*; 24) memorial with

address written on; 25) *bohyō* with ornate area in front; 26) *bohyō* with handwritten name; 27) memorial for two families; 28) sports-based memorial; 29) more personalized *bohyō*, showing signs of disrepair; 30) partly renewed *bohyō*; 31) Ken Miyajima's *bohyō*; 32) many *bohyō* and a crucifix in the area around Sugeno-sawa; 33) memorial for two foreigners; 34) a large stone memorial; 35) *bohyō* with no name visible on it; 36) Buddhist memorial stone; 37) flower, drink and cigarette left by a *bohyō*; 38) the faded original *bohyō*, next to the original metal post and tag, together with a stone *bohyō*; 39) rice left in front of the stone *bohyō* which has a stone baseball cap on it; 40) a magazine left in front of a *bohyō*; 41) a recent baby photo left by the *bohyō*; 42) 'One for all, all for one' – Keiko Kawakami's family's *bohyō*; 43) a tea cup left by the memorial; 44) a photo built into the *bohyō*; 45) a shrine; 46) clothing left on a post next to the *bohyō*; 47) a large memorial with a toy JAL plane left on it; 48) a selection of food left; 49) text discussing the causing of the crash left in front of the *bohyō*; 50) ornamental flowers and a ceramic dog left by the *bohyō*; 51) a Disney-style dwarf; 52) Buddhist stone memorial for one victim next to a more conventional wooden *bohyō* with its original metal post next to it.

particularly apparent around Sugeno-sawa, where most of the victims were found. Sugeno-sawa is also where the four survivors were found, but there are no memorials of any sort to this effect.

The final part of the 'Osutaka Pilgrimage' takes place at Irei-no-Sono (see Figure 6.9). The 11 metre-high main towers of Irei-no-Sono were designed by Tomihisa Handa and the site was constructed in time for the first anniversary.[105] Behind the towers the names of most of the victims are listed (see Chapter 7). In the centre, built into the hillside, is a crypt which houses the cremated remains that could not be identified. As Irei-no-Sono is hidden by a hillside in the village and there is nothing about the name which particularly connects it with JL123, those driving through Ueno-mura are unlikely to be drawn to the site by the signs for it. There are two back-to-back ceremonies held there on 12 August starting at 18:00. Many are dressed in black, with representatives from JAL often being in full morning suits, and some local people dressed in their work uniforms. However, there are many who are dressed more casually, probably due to them having come direct from Osutaka-no-One or because of the journey they have ahead of them.

The first of the ceremonies sees families, people from the village, JAL and others place flowers in front of the memorial tower. There are also a number of short speeches and a minute's silence. In the second ceremony, families light 520 candles, placed around the memorial, to remember the victims. A minute's silence is then held to coincide with the time of the crash. Although the ceremony is very poignant due to the sight of the 520 candles and the minute's silence coinciding with the time of the crash, some *izoku* say that this is the part of the pilgrimage which has the least meaning for them.[106] The key reason for this appears to be the lack of meaning of the location itself. For *tōrōnagashi*, the key meaning comes from the river, and in the case of Osutaka-no-One it is the *gemba*, the actual place where the crash happened. Irei-no-Sono, on the other hand, is on the site of a memorial people from Ueno-mura who died during wars,[y] and has no direct, original connection with JL123. But Irei-no-Sono is part of the pilgrimage today, and for those who cannot get to Osutaka-no-One, it is an important symbolic reminder of the events of 12 August 1985. Once the second ceremony ends, most leave Ueno-mura and few seem to stay that night. However, the last to leave are the representatives of JAL, including the company president, who remain in a deep bow until others have left.

Irei-no-Sono is significantly easier to visit than Osutaka-no-One and although it is not obvious what Irei-no-Sono is to the casual passer-by, for those wanting to pay respects to those who died on JL123 it is a relatively accessible location. It is not uncommon to see non-*izoku* there (see Chapter 8). Irei-no-Sono is run by the Irei-no-Sono Zaidan (Irei-no-Sono Foundation). The foundation was set up with a fund of ¥500m in 1986, with further sums of ¥300m, ¥6m and ¥5m added in 1988, in 2009 and 2010 respectively. As well as covering some of the expenses related to Osutaka-no-One and Irei-no-Sono (see Table 6.1), yields from the funds are used for the memorial services.[107]

Satoshi Ogawa 小川哲 *(41)* · *Masako Ogawa* 小川昌子 *(43)* · *Chisako Ogawa* 小川知佐子 *(8)*

Figure 6.9 Irei-no-Sono

Photographs by the author. a) Irei-no-Sono; b, c and d) statues around the Irei-no-Sono site; e) former Ueno-mura mayor, Takeo Kurosawa, aged 95, arrives for the memorial service in 2009; f) *izoku* line up to place flowers in front of the Irei-no-Sono towers in the first ceremony; g) *izoku* light 520 candles during the second ceremony; h) candles around the names of the victims at Irei-no-Sono, with red candles up the steps to the crypt where the cremated unidentified remains are; i) *izoku* place flowers and drinks on the names of loved ones and then take commemorative photographs.

Table 6.1 Cost of work done by Ueno-mura and Irei-no-Sono Zaidan

	Costs borne by Ueno-mura (¥)	*Costs borne by Irei-no-Sono Zaidan (¥)*	*Total (¥)*
Construction of the roads to Osutaka-no-One	430,728,000		430,728,000
Construction of road surface to Osutaka-no-One	5,768,000	38,712,000	44,480,000
Tree planting at Osutaka-no-One	10,240,000		10,240,000
Handrails at Osutaka-no-One		2,060,000	2,060,000
Construction of toilet building at Osutaka-no-One		2,230,000	2,230,000
Construction of car park at Osutaka-no-One		2,850,000	2,850,000
Construction of toilets at Osutaka-no-One		1,963,000	1,963,000
Osutaka-no-One costs sub-total	446,736,000	47,815,000	494,551,000
Construction of the road to Irei-no-Sono	47,580,000		47,580,000
Acquisition of land for Irei-no-Sono	30,250,000		30,250,000
Preparation of land for Irei-no-Sono	60,888,000		60,888,000
Construction of crypt and tower		231,800,000	231,800,000
Sewerage and administrative building costs		31,260,000	31,260,000
Rest-building construction costs		13,450,000	13,450,000
Raising trees and plants		22,000,000	22,000,000
Flower beds		3,605,000	3,605,000
Publication of Chinkon-no-Shiori		30,000,000	30,000,000
Publication of Flowers of Osutaka		14,177,000	14,177,000
Irei-no-Sono costs sub-total	138,718,000	346,292,000	485,010,000
Total	585,454,000	394,107,000	979,561,000

Source: Table by the author based on information on display at Irei-no-Sono.

The site itself contains a small museum displaying books, photographs and other materials – including many Buddhist artefacts. There is also a small rest area where an *omamori* (amulet) can be bought for 'traffic safety'. Over 70 per cent of Japanese use *omamori* at some time,[108] and, like with *ofuda* (talisman), are as much an indication of the individual's commitment to drive safely, for example, as a belief that *kami* will protect them from danger and so they can continue to go about their activities.[109] As such these items should not be seen as mere 'good luck charms' for if 'luck is a matter of chance, rather than the result of calculated deliberation, then a good luck charm is not about luck at all, since it symbolizes two kinds of causes that work co-operatively together: human effort and divine help'.[110] The number of amulets relating to travel being sold in Japan went up in 1985 and 1986 and is linked to the JL123 crash. For in 'the early 1980s it was quite rare to come across talismans, amulets, deities, or institutions that catered specifically to' air travel.[111] Similarly there is a place for people to leave prayers on *tanzaku*, as would be found at a Shintō Shrine, and also at Osutaka-no-One. Most of the messages left relate to travel safety generally or to the memory of those who died on JL123. Other messages call for help with examinations and sporting performances and have no obvious connection to JL123. This is comparable to the *ofuda* of pilgrims on the Shikoku Pilgrimage, for example.[112] As well as maintaining the site, over the years Irei-no-Sono Zaidan has also produced two publications.[z] Observing people going to Irei-no-Sono, whether prior to the official ceremonies (when it is likely to be *izoku* attending) or on other days (when those I observed were non-*izoku*), suggests that there is no consistency in how people behave. If one stops to pray before the steps leading up to the tower, then others around them will often do the same. However, many will not stop to pray.

The number of *izoku* who visit Osutaka-no-one and Irei-no-Sono has varied greatly over the years (see Figure 6.10). The graph is misleading in some ways due to the large number of people (1,785) who attended the first anniversary compared with subsequent years. In every year after 1986 more *izoku* visit Osutaka-no-One than attend the Irei-no-Sono ceremonies on the anniversary. Part of this may be due to the lack of connection with Irei-no-Sono. Another reason may be simply due to logistical reasons. Although Ueno-mura is easier to reach now than in the past, leaving the village at just after 19:00 makes it hard to get back to Tōkyō, let alone Kansai, before late that day. However, as mentioned above, one also has to remember there are others who are only attending the Irei-no-Sono ceremonies since the climb to Osutaka-no-One has become too difficult physically for them. Given that the average number per *izoku* group climbing to Osutaka-no-One does not vary greatly, being about 3.5, it would seem that perhaps once one family member cannot go, all stop attending.

Figure 6.10 includes a trend line for the number who climb to Osutaka-no-One on the anniversary. One can see that the line indicates that overall the number is not varying greatly. However, there are a number of years which are above the trend line. Of these 1987, 1991, 1997 and 2007 were significant Buddhist death anniversaries. For example, 1991 was the seventh anniversary using the Buddhist system of counting whereby 1985 is counted as year 1. As many of the families

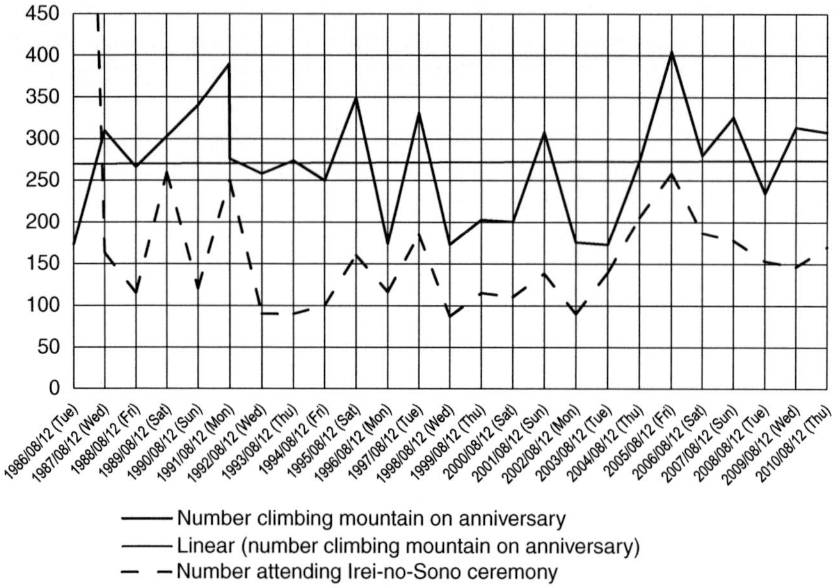

Figure 6.10 Izoku visits to Osutaka-no-One and Irei-no-Sono

Figure by the author, based on information from JAL and newspapers on 13 August each year. Note that in 1991 the Irei-no-Sono event was held on 4 August. This was due to it being the seventh Anniversary (using the Buddhist counting system) and the expectation that many families would have chose to go to the family grave locally on 12 August for Obon. Consequently there were a greater number of people visiting Osutaka-no-One on this date than the 12 August.

would have chosen to conduct the ritual at their home on 14 August, the closest Sunday to 12 August, an alternative day for them to show their respects at Irei-no-Sono was organized for 4 August,[113] but it is possible to see that on both 4 and 12 August the number that climbed was above the trend line. 1995 was significant due to being the tenth anniversary (using a standard calendar counting system). That Kōbe and the Kansai region was struck by the Great Hanshin Earthquake earlier that year may also have impacted numbers. 2005 was the twentieth anniversary. 2009 may have seen an increase due to the problems of visiting the site in 2008 and the construction of the new steps and barriers in that year (see above). Similarly, 2006 was the first year that the new car park and shorter walking route had been opened (see above). 2010 was the twenty-fifth anniversary. It would also appear as though the day of the week has some impact, with attendance on weekdays generally lower that at the weekend. Although it may appear surprising that the figure for 1986 is so low, officially climbing was not permitted that anniversary due to a typhoon, but many climbed regardless.[114]

It is finally worth noting that one can usually see at least one person, a priest himself, dressed in traditional pilgrims ware during the anniversary events.

According to Reader,[115] the key aspects of traditional pilgrim's dress is white clothing, such wrap-around *hakui*, which symbolizes a burial shroud. He also has a bamboo hat, *kasa*, which protects against the weather and symbolizes being in a coffin. A pilgrim also takes a staff, *tsue*, which is useful when hiking up and down hills and the steep flights of steps that lead up to many of the temples, and is symbolic of a *bohyō*. *Tsue*, made by local school children, including those in Ueno-mura and Fujioka, are also available at the car park at Osutaka-no-One for anyone to use when visiting the site. These walking sticks have messages on them and the name and class of the student who made it.

Although my discussion of the Osutaka Pilgrimage has focused upon 11 and 12 August, it is important to stress that many *izoku* will also visit Osutaka-no-One more than once during the year. Officially, the crash site can only be accessed between April and November each year because of the harsh winter conditions in the area. The number of *izoku* visiting the site each year has dropped, which probably reflects the fact many are finding the climb too hard as they get older. However, there are still many who will go to the site several times a year. Kuniko Miyajima, for example, has been to the site over 150 times in total.[116]

Conclusions

In terms of memorialization, JL123 shows that there is a shifting of the 'space' where memorialization takes place. Although the ashes of the dead are contained within the family grave, and for some this is the only place where memorialization takes place, for many the crash site (Osutaka-no-One) and the memorial in Ueno-mura (Irei-no-Sono) have become an additional space where their loved ones can be remembered. That the deaths were 'untimely' allows for this additional space to be created, but, there is more to it than that. Many others suffer 'untimely' deaths everyday in hospital from illnesses, yet memorialization such as that found for JL123 is unlikely to be found at hospitals. That so many died in a single incident, providing the potential for different families to come together to remember, is perhaps key in this respect. That the bodies were scattered across a wide area may also be significant as it allows for greater connection with their own loved one through individual *bohyō* rather than a single, shared memorial.

But Osutaka-no-One is not static. The number of memorials around Shōkon-no-Hi has changed over the years. *Bohyō* are changed from time to time, while those that are not changed, like many of the *sotoba*, will continue to deteriorate. Items left by *bohyō* and in the huts on Osutaka-no-One also change. Furthermore, there are natural changes due to the seasons and the way in which the vegetation at the site has grown back. The 8/12 Renrakukai also often look to change the way in which the get-together in front of Shōkon-no-Hi is done, whether it be through the releasing of balloons or using small windmills (see Figure 6.6i). Reader comments on how the Shikoku pilgrimage is a 'moving text that is constantly being rewritten by its participants',[117] in many respects this is also true of the Osutaka Pilgrimage too.

Although it is possible to draw some conclusions from what happened in 1985 and how it is remembered today, one has to be careful; with over 400 families, there is no defined response to the death and disaster in Japan. There are families who do not go to the crash site or memorials as they do not believe that the victim's spirit is there.[118] Yet, for many families it is important to do so. Building upon the ideas of Emmanuel Marx who suggested that pilgrimages reaffirm existing social order,[119] and Arnold van Gennep's study of initiation rites, Turner coined the concept of 'communitas', whereby groups create a new, temporary order.[120] This concept has been used by anthropologists ever since in studying groups of pilgrims. Although many will point to the weakness of this concept, most academics tend to still feel a need to refer back to it. Reader points out how many Shikoku pilgrims will not talk to each other when walking together as they get lost in their own thoughts.[121] Similarly, Aoki points out that although pilgrims to Ontake may travel together as a group, the reasons for travelling may be very different.[122] Externally the appearance is of togetherness, but in reality it is a false image in many respects. In many ways I find observing the *izoku* in Ueno-mura like I imagine how some would view a football crowd. From a distance there appears to be commonality in what is being done, said or sung and even worn, but differences can be seen upon closer observation. Not all football fans wear replica shirts, and those that do are unlikely to be all wearing the kit from a single season. Some will be singing one song, others another. Some will be shouting particular player's praises, while others will be more critical. There will be varied ideas about tactics. But in the end they are all still supporting the same team. The *izoku* are united in many respects in relation to JL123, but there is a variety of views and ways in which they conduct themselves, as this chapter has highlighted.

Thanks to the improvement of accommodation facilities in the village over the years, one can see a real bond between many of the families, and others, and that they look forward to getting together with these friends there again the next year. This is something seen among pilgrims more generally.[123] Indeed, for some families the pilgrimage to Ueno-mura has brought the family closer together and they see this as the positive development that has come from the tragedy.[124] Of course there is bereavement and sadness, but that has become less visible over the years, and one hears more laughter and sounds of children playing than the sound of people crying. It is also noticeable that the *izoku* are quite tactile compared to many Japanese and that they do not restrict themselves to bowing. I often see them holding each other's hands when talking to each other. It is also not uncommon for those coming down from Osutaka-no-One to exchange greetings with those going up – the latter saying something like '*otsukaresama*' and the former saying '*gambatte kudasai*' (keep going) – although such exchanges may be common to climbing in mountains more generally and not restricted to this particular location.

Watching the memorial events it is clear that JL123 has become about more than simply this crash itself. The *tōrōnagashi* is very explicit about this. One of the lanterns contains details of other tragedies being remembered such as the Shigaraki Kōgen rail crash in 1991, the China Airlines flight CI140 in 1994, and the Fukuchiyama line derailment in 2005. Representatives of associations connected to these accidents are often also present. I would suggest the fact that the journey to Ueno-mura and particularly Osutaka-no-One is so difficult is a vital ingredient in the way in which 'the Osutaka Pilgrimage' has developed. That Osutaka-no-One is a mountain, now considered by some to be a *seichi* (sacred place), is also highly significant. The timing is also significant, as 12 August is when many get ready to return to ancestral homes to see family members and remember those who have died. It also fits into a pattern, with Japan remembering the bombing of Hiroshima on the 6 August, the bombing of Nagasaki three days later, and then the end of the war three days after 12 August. In 1985 this was of particular focus due to it being the fortieth anniversary and Prime Minister Nakasone making the first visit as a serving Prime Minister to the controversial Yasukuni Shrine on 15 August. The crash also comes at a time of year when many people also remembering their childhood because of the Kōshien high school Summer Baseball Tournament, where some JL123 victims were travelling to, going on. That the crash of a plane, a 747 – a symbol of modernity itself – killed many *tanshinfunin*, who were slaves to the economic machine, and others returning home for Obon, came at a time when many Japanese were concerned that Japan's modernization was coming at the expense of traditions, which was leading to 'retro booms',[125] is probably not without significance. Furthermore, many of those who lost loved ones on JL123, as well as others in JAL and Ueno-mura, were likely to have experienced the upsurge in *ireisai* (memorial ceremonies) following World War II, as those left behind wanted to consol the soul of those who had died as well as dealing with their own guilt for the person who had died.[126] These ceremonies may have helped condition what expected behaviour would be for the post-war generation.

As discussed, ancestral worship is very important in Japan, but in relation to JL123 we are yet to reach the point where victims have become ancestors as this tends to be on the thirty-third or fiftieth anniversary. In the case of the ANA flight NH58 crash, the company used the thirty-third anniversary as the time to draw the official memorials to a close in 2003,[127] although families and representatives of ANA continue to go to the site for the anniversary.[128] Whether JAL will attempt the same for the ceremony at Irei-no-Sono is yet to be seen.

The Osutaka Pilgrimage reflects many aspects of pilgrimage in Japan. Although it is not circular (especially due to the need to go up and down a single road that links central Ueno-mura with Osutaka-no-One), it does contain a number of points which have to be visited. Although people's ideas about what happens after death vary, the rituals related to the death of a loved one are often largely the same, but with JL123 there are some differences. Although there is commonality in much of what each family does in Ueno-mura, there are subtle differences and we must not forgot that there are many more *izoku* who do not go on the Osutaka Pilgrimage.

There is arguably an additional part of the Osutaka Pilgrimage which I have not covered in this chapter, and that is the Safety Promotion Center at JAL (see Chapter 8). I have not included it as part of the pilgrimage as some *izoku* travel there quite separately from travelling to Ueno-mura. Additionally, what is there and what its purpose is best understood once having discussed the non-*izoku* visitors to Ueno-mura in Chapter 8.

One thing that I have not explored is why some *izoku* do go on the Osutaka Pilgrimage. At one level the reasons seem obvious and I have discussed the issues in this chapter, but there could be different motivating factors for each *izoku*. One aspect which cannot be overlooked is, what Reader describes as 'The sense that "I had to come"', for 'even though ostensibly voluntary in nature, [it] may contain a degree of compulsoriness'.[129] Reader points out that 'This compulsory volition in pilgrimage is not always conditioned by individual and inner drives and needs, however, but by social and cultural ones as well'.[130] In relation to the Osutaka Pilgrimage, what would the reaction of the other *izoku* and media be if Kuniko Miyajima did not go on the Osutaka Pilgrimage, for example? I expect there are also family members who do the Osutaka Pilgrimage because of the expectations of other family members rather than having their own desire to do so, just as Reader has found with the Shikoku Pilgrimage.[131] But we should not underestimate the commitment of those who do go. Although Ueno-mura and Osutaka-no-One is easier to access today than in the past, a two-day trip there can easily cost in excess of ¥40,000 per person if coming from Kansai.

Part IV

Narratives about JL123

7 The media and JL123

> Major lives and minor lives. Important lives and lives which aren't… For the people of the media the people who died in the JAL crash were extremely important lives, weren't they?[a]
>
> *Hisashi Ōshima 大島久 (also known as Kyū Sakamoto 坂本九) (43)*

Let us return to the evening of 12 August 1985. By 19:30 the news of JL123's disappearance was being flashed on TV screens across the country. Meanwhile, away from the public eye, the media machine was going into overdrive as it tried to cope with one of the biggest stories in the post-war period. Given the scale of the crash it is not surprising that it became such a media event. This chapter will look at the way in which the crash was covered by various media in 1985 and also look at the coverage of the Osutaka Pilgrimage.

This chapter addresses three main issues. First, how much uniformity is there in the Japanese media? Throughout the chapter I will highlight the similarities and differences between the various types of media (newspapers, *shūkanshi* (weekly magazines) and TV media) in addition to the similarities and differences between national and local coverage of the crash. Second, this chapter discusses issues relating to privacy, with particular reference to the printing of the passenger list, the interviewing of the survivors, photographing the dead and interviewing *izoku*. Third, this chapter discusses the media's role in creating a memory of an event, which is a major theme of this book.

Different media and the role of the media

Before looking specifically at the reporting of JL123, let us begin by considering the different types of media in Japan (television, newspapers (national and local) and magazines) and the way in which they operate.

The main national newspapers have a close affiliation with national TV stations also (see Table 7.1). Newspapers are officially allowed to own ten per cent of the TV stations, but it is thought that through the use of other affiliated companies that the figure 'exceeds 50 per cent'.[1] As well as the main national and local newspapers, there are also some news agencies, such as *Kyōdo News* and *Jiji Press*, which collect news and then sell them to other media organizations, particularly

Table 7.1 Japanese newspapers and affiliated TV stations (ranked by morning sales)

	Newspaper	*Morning sales*	*Evening sales*	*Affiliated TV station*
1	*Yomiuri Shimbun*	10,016,735	3,570,650	Nippon TV
2	*Asahi Shimbun*	7,955,595	3,120,302	TV Asahi
3	*Mainichi Shimbun*	3,593,867	1,241,227	TBS
4	*Nikkei Shimbun*	3,050,277	1,613,701	TV Tōkyō
5	*Sankei Shimbun*	1,756,410	556,887	Fuji TV

Table by the author.

Sources: Akhavan-Majid 1990; *Yomiuri Shimbun* 2010, *Asahi Shimbun* 2010, *Mainichi Shimbun* 2010, *Nikkei Shimbun* 2010, *Sankei Shimbun* 2010; . Data is the average sales between January and June 2010 for the *Yomiuri*, *Asahi* and *Mainichi*; July to December 2009 average for the *Nikkei*; January to June 2009 average for the *Sankei*.

local newspapers. One important point to note with the data in Table 7.1 is that sales do not necessarily equate to readership. About 95 per cent of newspaper sales are done by subscription and home delivery.[2] The tactics used by the newspaper salesmen trying to encourage someone to take out a subscription can range from the offers of introductory gifts and prices to more emotional approaches.[b] This could mean that some people end up with a subscription for a paper that is barely read. I have come across many Japanese who get the *Nikkei Shimbun*, for example, but they only read a few selected pages of it and get another paper for their main news. Others I have come across only use the newspaper to get the TV listings and remain informed through viewing the TV news.

A particular feature of the media in Japan is the use of *kisha kurabu* (press clubs). Within major companies and public organizations there will be a *kisha kurabu* with representatives from the main national newspapers and news agencies. At the time of the JL123 there was no foreign representation.[3] The companies and organizations provide space, access to telephones and sometimes other ICT facilities to the journalists,[4] and these journalists will then be provided with briefings when appropriate.[5] Should journalists write stories that do not meet with the approval of the company or organization, they could find that their membership of that particular club is revoked.[6] Consequently, one would expect that the system leads to conservatism, as journalists will not seek to investigate on behalf of the public and there will be a tendency for journalists of the various newspapers to barely rewrite what they are provided, so the content of newspapers becomes very similar. The concern is that the Japanese media have become so subservient to those it gets its information from, as well as the sponsors who pay for advertising in their papers, that they have ignored what their role should be. Indeed, mainstream media (newspaper and televisionc) may be so subservient to their masters that it might not be possible to divorce the media from the state.

Kunio Ōtake 大竹国雄 *(44) · Yoshinori Ōyuki* 大幸義典 *(43) · Setsuko Ōyuki* 大幸節子 *(36)*

Whilst these concerns surround television and main national and local newspapers, other sections of the media have greater freedom. There are some so-called sports newspapers, which focus on sports but also contain some national news. Their content is perhaps more tabloid-like than other newspapers. However, the main tabloid news in Japan is provided by weekly and monthly magazines – *shūkanshi* and *gekkanshi* – which are not permitted to join the *kisha kurabu*. Although many Japanese do not believe what is written in these magazines, their influence is not insignificant and is not restricted to just their contents, with adverts, featuring the headlines from the articles in them, showing up 'everywhere'.[7]

We need to remember that, for the media, the reporting of an event such as the JL123 crash is a job; a particular job that can help their business. As discussed in Chapter 2, many people have a fear of flying and perhaps as a consequence of this there tends to be great interest when there is a major plane crash, but the level of interest and reporting is totally out of proportion to issues relating to general safety.[8]

Reporting the crash

Let us now turn to focus on the reporting of the crash itself. As well as interviewing journalists and looking at the original newspapers, there are other materials which have been very instructive. In particular, the way in which the first 24 hours of the JL123 crash was covered by *Asahi Shimbun* was detailed in a book written by the Asahi Shimbun Shakaibu. As noted in Chapter 4, the *Asahi Shimbun*'s helicopter was among the first to reach the crash site and the photograph taken then appeared on the front page of their newspaper the next morning (see Figure 7.1). The main headlines (going in a clockwise direction from the top) read 'JAL jet carrying 524 crashes', 'Inferno at the Nagano/Gunma border', 'Haneda plane bound for Ōsaka: serious damage, all souls lost?', 'Straight after contact saying "door broken"'. However if one looks at the map on the paper, the point of crash does not appear to fit with what the *Asahi Shimbun* itself thought was the crash site,[9] and is perhaps an indication of the confusion that existed during that night about the location (see Chapter 4) and a reluctance not to print information contrary to what was being said by the authorities.

Having looked at the front page of the *Asahi Shimbun*, let us look at that of the *Jōmō Shimbun*, a local newspaper based in Maebashi and primarily serving Gunma prefecture (see Figure 7.2). In other words, although this would not be confirmed until after dawn on the 13 August, the crash happened within their area. Looking at the *Jōmō Shimbun*'s front page, many of the headlines are similar to those in the *Asahi Shimbun*; reading clockwise from the top they are 'JAL jumbo crashes, inferno', 'In mountains on the Nagano/Gunma border', 'Have all 524 passengers and crew been lost?', 'The world's largest plane disaster'. However, the next headline (which has been partly distorted in Figure 7.2) reads 'Mr Takeshita (father of a player in Nōdaini) from Takasaki also [on board]'. Nōdaini is the nickname of Tōkyō Nōgyō Daigaku Daini Kōtōgakkō, the Second High School, Tōkyō University of Agriculture, which is located in Takasaki. In 1985 it was

朝日新聞

524人乗り日航機墜落

長野・群馬県境で炎上

羽田発大阪行き 大破、全員絶望か

「ドア破損」と連絡直後

河本特命相が辞表

Figure 7.1 National newspaper headlines 13 August 1985

Picture courtesy of *Asahi Shimbun*.

THE JOMO SHIMBUN

上毛新聞

日航ジャンボ機墜落、炎上

長野・群馬県境の山中

乗客ら524人絶望か

世界最大の航空機惨事

高崎の[redacted]さん も

ドキュメント

断片情報にいらだちの声

県警に対策本部

便利さ、ご存知ですか。

Figure 7.2 Local newspaper headlines 13 August 1985

Courtesy of *Jōmō Shimbun*. The name of the passenger and his family in the headline and the story has been deleted by request of *Jōmō Shimbun*.

representing Gunma prefecture in the summer high school baseball tournament at Kōshien. This headline points to two important points. First, issues of privacy might be different in Japan to what is seen in many other societies (see next section), and second, that the newspaper was keen to put the focus of the story on the local issues. Although it was not yet certain that the crash site was within Gunma (and the map indicates the location being in Nagano), they were certain that there was a passenger from Gunma on board. Comparing the two front pages, *Asahi Shimbun* had a photograph of the actual crash site, whereas the smaller resourced *Jōmō Shimbun*'s main image is a picture of another JAL plane with an arrow pointing to the door which was reported as being broken by the cockpit crew and so suspected to be related to the cause of the crash.

Meanwhile, foreign journalists were also struggling to get hold of information. William Horsley was the BBC's correspondent in Tōkyō at the time of the JL123 crash and remembers spending time flicking between the TV channels trying to collate information for reporting back to London.[10] During the first few hours there was much confusion over the location of the crash site. This created a problem for the Japanese media due to printing deadlines for newspapers and most television stations in those days would go off air (see below). However, in the UK, for example, when news of the crash broke it was still 11:30 and so there was time for the popular afternoon newspapers to gather information, for the TV news to find out more details for the main evening bulletins, and for the national newspapers to start preparing headlines for the next day's papers. But with very limited, accurate information available in Japan, getting the necessary information for readers and viewers back in the UK was very difficult.[11]

Although we now live in a world of 24-hour television news, in the 1980s this was largely unheard of, and most TV stations shut down at night. When the news of JL123's disappearance first broke, the TV stations posted messages across screens, as is done when announcing other major events such as an earthquake and whether there is any tsunami risk. This was the first that any of the public, including relatvies of those on board JL123, came to know of what had happened. Whilst some TV stations continued with their normal TV schedule with the inclusion of some news flashes, others stopped their normal broadcast and focused upon the events relating to JL123. TV ratings surged to a peak of over 75 per cent for the two hours between 20:00 and 22:00, ten per cent higher than normal, in the Kantō region.[12] Due to the scale of the story, not all TV channels ended their broadcasts in the normal way that night. Whilst the TV stations could provide updates throughout the night, time was critical for the newspapers so they could be printed and delivered to their subscribers and shops for selling. Consequently, every night the papers are 'put to bed' at a particular time in order to meet this deadline, but with information changing constantly through the evening, as well as papers being keen to include detailed information on the passenger list (see below), the normal deadline was a problem. So the national newspapers agreed with each other to a later-than-normal print run.[13] Such an agreement seems quite unimaginable if one thinks of the competitive advantage a paper could get from having its paper ready on time with exclusive information or pictures, and is perhaps both a reflection of how extraordinary this event was and also the difference in the nature of the media business in Japan.

On the morning of the 13 August, the TV camera helicopters had the advantage of being able to quickly locate the crash site from the air and so these images then dominated. Viewing figures for that morning were around 20 per cent higher than normal, reaching a peak of over 60 per cent, in the Kantō region.[14] Meanwhile journalists were homing in on the crash site by ground and some reached the crash site before the SAR teams.[15] Once there they began to take photographs and write about what they saw around them, leaving opened cases of camera films littering the crash site in the process,[16] and so further contaminating the site. Many of these journalists were ill-prepared for climbing to the site and were wearing the wrong types of shoes and clothing, such as neckties, smart shoes or even sandals.[17] There were even cases of SAR teams having to go and find missing journalists in the mountains rather than focusing their energies on the crash site.[18] TV camera crews later arrived at the site and they had larger, heavier equipment. When the crews left the site for a time sometimes they even left their equipment, possibly worth more than ¥10m, with a note to say 'We'll be back to get it later. Thanks'.[19] In amongst the chaos of the reporting, some of the most iconic pictures of the post-war period were taken – such as that of a survivor being taken away from the site (see Figure 4.4) and of the

Figure 7.3 Biggest stories of the post-war period

Picture courtesy of *Nikkei Shimbun*.

site itself. In an article run in 1995, the *Nikkei Shimbun* described the JL123 crash as one of the biggest stories of the post-war period, and a photograph similar to that in Figure 7.3 was one of only seven images in the article. Fuji TV's iconic images taken of one of the survivors' rescue helped it to win a Japanese Newspaper Association award in 1985.[20]

The passenger and crew list

One of the things which newspapers were trying to do for inclusion in their morning editions of 13 August was a passenger and crew list. The TV news had already announced the passenger list, but the newspapers also attempted to include details, such as having the name written in the correct *kanji* (the passenger list was initially provided in *katakana* format only),[d] their age, where they were from, their home address and why they were on the flight. It is worth reflecting upon the logistical problems in doing it due to the numbers involved. The *Asahi Shimbun*, for example, assigned 30 people to the task,[21] the results of which are discussed below. Additional details, such as photographs, were added over the coming days, and articles began to resemble those done for *kamikaze* pilots during the war.[22]

So why were the newspapers printing the passenger and crew list? In many respects, this question appears to be asking the wrong thing. With a lack of privacy laws and no concept of next of kin being informed first, the question should be why *not* print this list? There is clearly a desire for immediate information, even if it was at the expense of accuracy for some names on the original list given out by JAL were incorrect.[23] I was living near Nagoya in 1994 when flight CI140 crashed. I remember being shocked at seeing the list of passengers appearing on TV within a few hours as I expected next of kin to be informed directly first. Yet, of those who accessed one of my web pages after doing an internet search (see Chapter 8), let alone emails that I get relating to my research, one of the main requests is for information relating to the passenger list. Clearly there is an expectation that such information should be in the public domain at some stage. For some, the list was a means to try to confirm whether someone they knew was on the flight, although I have come across some who saw the JL123 passenger list on the TV news and thought that someone they knew was on the flight only to discover that in fact it was somebody else with the same details. However, in the longer term, based on some sites I have seen relating to other tragedies, it seems the passenger and crew list is used so that their names will not be forgotten and so similar tragedies will not be repeated. This is in-keeping with one of the aims of the 8/12 Renrakukai. That there is a need to remind people of the size and nature of tragedies was demonstrated by findings in 2010 that children even within Kōbe could not comprehend the scale of the Great Hanshin Earthquake.[24]

Providing a list of names can serve a purpose. I wanted to include the passenger list in this book, but the way to do this vexed me for some time. Although several pages of names in an appendix would still undoubtedly have some impact,

Gab Soon Choi 崔甲順 (also known as Fujie Ihara· 井原富士恵) (45)

as it does in the Asahi Shimbun Shakaibu's book, I was concerned that many readers would not look at this part of the book and so the impact would be lost. I finally decided to include the names in the footer so that there would be a constant reminder of the human tragedy involved (as discussed in Chapter 1). That it takes so many pages to list all the names is significant and this became more evident to me when I went to see other memorials (see Figure 7.4). No names are listed at the memorial for the victims of the Mikawashima railway accident, but there are names are listed for the over 6,000 victims of the Great Hanshin Earthquake. That Japan may be changing in its approach to memorialization and the use of names can be seen by the fact that even four years after the accident, there was no decision about the design of the Fukuchiyama Line derailment memorial, although common elements to the memorialisation of JL123 – as seen in Figure 6.8 – can be seen. However, seeing the memorial for another aviation accident, the Air France AF4590 Concorde crash, provided a more direct comparison with JL123. There are no names listed there. The reasons for this and the design of the memorial itself are not relevant to this study, but it is suffice to say that I found the

Figure 7.4 Other memorials

Photographs by the author. a) a statue of Kannon for the victims of the 1962 Mikawashima rail accident at a nearby temple, Jōshō-ji; b) one of the memorials for the Great Hanshin Earthquake – this underground one, with a clear ceiling so that a water feature above can be seen, has the names of the victims listed around it; c) the provisional memorials in 2009 for the victims of the 2005 Fukuchiyama line derailment in Amagasaki; d) the memorial for the victims of Air France flight AF4590 Concorde memorial in Gonesse, France.

Ji Hye Park 朴知恵 (also known as Tomoe Ihara· 井原智恵) (14)

absence of names, more than its smaller and less impressive design, the most significance aspect as to why I felt its emotional impact is much less than that of Irei-no-Sono, which has most of the names listed.

Providing the passenger list is not without its problems. First, there are issues of privacy and data protection. That so many people search for the list is an indication of how little it is in the public domain compared with many other aircraft accidents. But, this may in part be due to it being an overwhelmingly Japanese tragedy. All but 22 of the 524 passengers and crew were Japanese. The passenger and crew list is available in a variety of forms in Japanese, and what is presented in this book is merely a reproduction of them rather than the exposing of information which has not been made public before.

But there are different versions of the list. First is the list included in Asahi Shimbun Shakaibu's book, which is an updated version of the ones that appeared in the *Asahi Shimbun* in 1985. Second, *Yomiuri Shimbun*'s evening edition on 24 August 1985 provided a detailed seating plan. This seating plan not only provided names but also details about which passengers and crew had been identified up to that point and how they had been identified. There is also a publication by Irei-no-Sono and which was available for purchase at Irei-no-Sono,[25] as well as the list of names at the Irei-no-Sono memorial itself. Finally, there is the memorial stone at Osutaka-no-One. Considering these five, there are differences in the list. For example, the list of names at Irei-no-Sono has 516 names not 520, whilst Irei-no-Sono's publication contains 506 names. The other three sources contain all the names – including the four survivors in the case of the two newspaper articles. The Irei-no-Sono list notes that one of the passengers was pregnant (see Chapter 5).

The order of the names varies with the lists. The *Asahi* list and Osutaka-no-One list are essentially alphabetical (using the Japanese phonetic alphabet). However, in both cases the foreigners are listed together at the end of the list of Japanese passengers, with the crew listed together after that. The Irei-no-Sono publication also keeps the foreigners largely listed separately and the crew at the end. However, the Japanese fiancée of one of the foreign victims is listed alongside his name rather than among the Japanese victims. This system is also used at the Irei-no-Sono memorial. Additionally, at the memorial, and in some cases in the publication, friends and relatives are listed together rather than using the Japanese alphabetic order. The crew are included in the Irei-no-Sono memorial list according to the Japanese alphabet, whereas the *Yomiuri* diagram keeps the crew listed together rather than showing where they were likely to have been located on the plane.

The Irei-no-Sono lists also reveal another issue with names which reflects a deeper issue in Japanese society. Four names appear which do not appear on the other lists. Only by reading the small text by one of these names at Irei-no-Sono can the discrepancy begin to be understood. The name by which one of them was listed on the passenger manifest is also provided in a smaller font. The reason for the two names would appear to be that these four individuals were *zainichi* – Japan-born Koreans. Although born in and residents of Japan, legally these people have been treated in many respects as foreigners in Japan.[26] This issue

Gi Jong Park 朴基宗 *(also known as Tatsuyoshi Arai* 新井立悦*) (32)*

has been a source of conflict for many and is beyond the scope of this study, but clearly the use of their Korean names at Irei-no-Sono is a reflection of the displeasure of some in using their Japanese names.

Having decided that I would include the passenger and crew list, the problem was the order of the list and whether to include all the names. In relation to the second point, I concluded that it would be most appropriate to include all the names – including the four survivors (whose names are written in bold). As noted in Chapter 3, some passengers might not have been sitting according to their seat allocation, and so I did not believe it appropriate to provide detailed information about the seating plan in this book or to list people according to the seating plan. I was also concerned that using the seating plan would mean that in some cases that family members would not be listed together. So I decided to use the Irei-so-Sono lists as a basis – i.e. keeping relations and friends together – but listing the names according to the English alphabet rather than the Japanese one. The main variation from many of the other lists is that although I thought it appropriate to keep the crew together, I decided to list these first as I believed that it is appropriate that the list be headed by the captain.

Another issue that I considered was how much information to provide with the passenger's name. Given the importance of age in Japanese society (see Chapter 8) and the particular impact that seeing a child's details listed is likely to have upon the reader, I decided to include the ages. However, this was not straightforward, as I found a number of inconsistencies between the sources. Where conflict was found I used the Irei-no-Sono publication as I believe this one had been more carefully researched and its contents checked with the *izoku*.

The media printing details such addresses did lead to problems for some *izoku*. For example, one told me that he was harassed by cults and such like after the crash which killed his parents and sister. He believes that they most likely got his details from the media.[27] As can be seen with the way in which Figure 7.2 has been reproduced, newspapers today have become much more concerned with privacy issues. Although rulings relating to protection of privacy have been around for many years in Japan,[28] it has taken further legislation, culminating in the Personal Information Protection Law in 2003, to force companies and organizations to reach a position which would be considered more comparable to that found in countries such as the UK. As a result, I doubt that we would see names being listed in the way it was done for JL123 again in Japan.

Many recent Japanese books about JL123 use pseudonyms or an English letter to refer to a victim. But surely there is a danger in doing this that the person's existence is being denied and they are being brushed under the carpet as though they were never born. Yet, for some reason, this happens less with famous people. Is this because, in the case of Kyū Sakamoto, it was not his real name anyway? Or is it merely acceptable to treat such high-profile people differently? I believe that the issue of privacy may have gone too far in Japan and lost its focus. It has lurched from one extreme to another; from being too open to being too closed. I find the media's openness in reporting JL123's passenger names before the next of kin had been notified distasteful, but the swing to not using passenger's real names is disrespectful to the victims themselves.

Getting the stories

Returning to the reporting in 1985, once survivors were found, this became the focus of attention for some of the media. However, as Fuji TV were on the scene to get pictures of the survivors being taken away and were keen to pass these images to its affiliated newspaper, *Sankei Shimbun*, the agreement to allow later print runs was terminated.[29] Some of the media set about trying to interview the survivors – who, in the case of the two adults were referred to by their family names, whilst the children were referred to by their first name and the suffix for (female) children, *chan*. The steps that some of the media went to included dressing up as doctors to try to access the hospital rooms and climbing around the outside of the hospital building.[30] Having seen the hospital I cannot comprehend how the desire to get a photograph from such an angle could possibly have warranted such extreme action, which could have easily led to a fatal accident. Given that the majority of sales are based on subscription, it is perhaps also surprising that papers were apparently taking steps to try and get an 'exclusive'. But it was not only at the survivors' hospitals where journalists were taking such extreme steps. There are also suggestions that some dressed up as SDF soldiers to access part of the crash site and as doctors to access the morgues.[31] However, large numbers of journalists hovered around the gyms and other buildings in Fujioka waiting for news releases and trying to picture and interview *izoku* (see Figure 7.5).

Many *izoku* did agree to be interviewed while they waited for news and to be asked to try to identify remains. However, whilst many journalists (estimated to be between 600 and 700)[32] were hunting for stories in Fujioka which would highlight the human tragedy, those at the crash site were presented with a story. On 15 August

Figure 7.5 The media in Fujioka

Picture courtesy of Peter Mathews.

Mr and Mrs Miyajima made their way to the crash site to search for their nine-year-old son, Ken, who had been travelling by himself for the first time in order to visit his grandparents to see a high school baseball game at Kōshien. The crash site was meant to be off limits to all except the SAR teams and journalists, but others did access the site, and some SDF soldiers made no attempt to stop *izoku* but instead wished them good luck in making the trek.[33] Many *izoku* took flowers, food and drink to place at the site in accordance with traditional remembrance activities (see Chapter 6). However, Mr and Mrs Miyajima firmly believed that their son was still alive and took a change of clothes with them. Despite seeing some of the images on the TV news, they believed that their son could be lost in the mountains somewhere, but on arrival at the site they saw the grim reality of the crash site for themselves and realized that there was no hope. Mrs Miyajima fell to her knees and wept saying 'sorry' and 'it must have been hot'. The media were there to capture this moment. The photograph (see Figure 7.6) appeared in the *Yomiuri Shimbun* the next day together with their story. Mrs Miyajima herself is unsure of the significance of the picture,[34] a colour copy of which is hanging in her lounge and also adorns the front cover of her 2010 book, but I suspect that it and the story it represents, is one of the reasons why she became the figurehead of the 8/12 Renrakukai and that Ken is one of the most well known victims (see Chapter 8).

Figure 7.6 Izoku at the crash site

Picture courtesy of the *Yomiuri Shimbun*.

Tetsuo Sakata 酒田哲雄 *(35)* · *Shigeko Sakurai* 桜井繁子 *(23)* · *Masako Satō* 佐藤マサ子 *(37)*

Reporting and visualization of death

Whilst the media was reporting about death and the photographs often implied death, on the whole there were not many images showing dead bodies or body parts. The newspapers and TV media focused on images of coffins in Fujioka or bodies wrapped up in blankets at Osutaka-no-One, but some magazines were far more graphic. This section discusses the use of these images. None of the images are included within this book, but internet addresses for a selection of them are provided. As sites showing these images are often removed, you may need to use a search engine such as www.archive.org to access them.

In the weeks after the crash, magazines such as *Emma, Focus* and *Friday* ran stories with graphic pictures of the scenes at the crash site. These weekly magazines regularly had combined sales of over half a million – and, unlike newspapers, more than 90 per cent are bought at shops, so sales are likely to be a good indication of readership level.[35] Although sales are small when compared with newspaper circulation, they are not insignificant. The pictures included in these magazines include SDF soldiers trying to reach limbs down from trees and other shots of limbs hanging from trees,[36] corpses lying with blankets on them,[37] burnt and mangled limbs and bodies,[38] the crash site and bodies wrapped up in blankets waiting to be transported off the mountain to the temporary morgues[39] and photographs of bodies inside coffins in the morgue itself.[40] These images raise a variety of questions. Why would people want to publish them? Why do some people want to see them? And why do some people – whether they be *izoku* or not – find the publication of these images so repulsive? But even these publications have limits. For example, one picture shows a child in a coffin but the face has been distorted so that it is not recognizable. Is this because the person would otherwise be identifiable, because they are in a coffin, or because they are a child?

We have become accustomed to seeing similar images of the aftermath of the bombings of Hiroshima and Nagasaki. We also see similar images in documentaries and films made about the bombing of Hiroshima in particular. The link between JL123 and Hiroshima was even made by one the magazines when it used the caption 'no more JAL 123' on one of the pictures in an apparent nod to the 'no more Hiroshimas' slogan. While some of the pictures of victims of Hiroshima and Nagasaki were of the living, viewers would probably realize that many would not live for much longer. Films or documentary dramas about Hiroshima can look incredibly realistic, but we console ourselves with the assumption that the dead are models and not real. But why do we find it easier to accept the images of victims of Hiroshima and Nagasaki in comparison to those of JL123? Is it because in the case of JL123, most victims were ultimately identifiable or had living relatives? Is it because permission had not been given for the images to be used? Is it because there is a feeling that these magazines are merely exploiting the JL123 victims rather than being educational? Is it because for many Japanese these victims are not considered 'dead' at this stage (see Chapter 5)? However, that these pictures were published and people bought the magazines is an indication that not all do find such pictures distasteful. I met one non-*izoku* at Osutaka-no-One who made a point of saying that these pictures do have value without me even raising the issue.

One of the photographers for *Focus*, Naonori Kohira, went on to publish a book of some his photographs. The book, *4/524*, takes its title from the number of survivors out of the total number of crew and passengers. This is a tall, thin, and, by Japanese standards, expensive book with limited text, of which most is in English. Although the title reflects the number of survivors, only 16 of the 51 pictures in the book are of the four survivors. Ten pictures are of the dead, presented in a section called '520/524'. All are of fully intact bodies. This may be as much due to the limited area the author appeared to have worked in at the crash site as much as due to issues of decency. Thirty-nine of the images show rescuers and 35 show a significant amount of plane wreckage. There are hardly any pictures focusing on personal effects, although one of these – of an ace of spades card – inspired a storyline in *Kuraimāzu Hai* (2008) (see Chapter 8), although not mentioned in the original novel.

Reporting JL123 in August 1985

Having discussed the imagery and some of the key ways in which the media covered the JL123 crash, let us now return to looking at the whether there are any significant differences between the way that newspapers wrote about the crash. To do this I have used the newspapers listed in Table 7.1. My analysis covered the newspapers dated 13–31 August, looking at the phrasing of headlines, the contents of articles, the use of photographs, when and what stories were written, which sections of the newspapers the stories were presented, and when the news about JL123 was reduced. Space prevents me from presenting all of the results here, but I will provide details which are indicative of my findings.

Table 7.2 lists the top headlines (*ichi-men-toppu-no-midashi*)[e] of each of the five main national newspapers in their morning edition of 13 August. All five papers used the word *tsuiraku* which means a plane 'crash', but four of the papers specifically mentioned that it was a JAL plane. Only the *Mainichi* did not mention that it was a JAL plane, and so readers would only be aware that it was a plane crash due to the use of the word *tsuiraku* rather than *jiko*, which would be used for more general crashes and accidents. Three of the five newspapers used exactly the same headline. However if we consider the side headline (*waki-no-midashi*) on the right-hand side of the paper we see greater diversity. All the papers also featured a picture of the crash site – three of them using the same image taken by an SDF helicopter, whilst the *Asahi* and *Yomiuri* used images taken from their own helicopters. Table 7.2 also notes the suggested location of the crash site in the side headline. Whilst four are relatively vague about the location, the *Yomiuri* suggests that the site is near Osutaka-yama, as discussed in Chapter 4. All five newspapers also had a map on their front page. Only the *Yomiuri*'s does not show the possible route of JL123 on the map but merely details some of the key place names on it. The *Asahi*'s map appears to suggest that the crash site is on the Nagano side of the prefectural border, despite the fact that their picture confirms that they had found

Sanae Satō 佐藤早苗 (39) · Yoshihiko Satō 佐藤喜彦 (12) · Kiyoe Satō 佐藤公由江 (10)

Table 7.2 Headlines on 13 and 14 August 1985

Newspaper	*13 August*		
	Top headline	*Side headline*	*Location of crash site*
Asahi Shimbun	JAL plane crashes with 524 people on board	Inferno on the Nagano/Gunma prefectural border	Nagano/Gunma prefectural borders
Mainichi Shimbun	Crash carrying 524 people	JAL plane crashes; Depart Haneda Destination Ōsaka	Nagano mountains
Nikkei Shimbun	JAL plane crashes, 524 people lost	Jumbo, wreckage in mountains	Nagano/Gunma prefectural borders
Sankei Shimbun	JAL plane crashes with 524 people on board	All passengers and crew lost	In mountains on Nagano/Gunma prefectural borders
Yomiuri Shimbun	JAL plane crashes with 524 people on board	Aviation history's biggest, unprecedented horrible accident	Near Osutaka-yama (Gunma/Nagano prefectural borders)

Newspaper	*14 August*		
	Top headline	*Side headline*	*Photograph(s)*
Asahi Shimbun	–	Part of the tail, found in Sagami Bay	Rescuers taking Keiko Kawakami out of the plane wreckage; picture of Hiroko Yoshizaki's face; picture of Mikiko Yoshizaki's face; picture of Yumi Ochiai's face
Mainichi Shimbun	About 80 bodies identified	Vertical stabilizer in Misaki waters	Keiko Kawakami being winched up with SDF soldier; retrieval of part of the tail
Nikkei Shimbun	One part of the tail discovered in Miura waters	Direct hit by the damaged door?	Retrieval of part of the tail
Sankei Shimbun	Recovery of upper part of the vertical stabilizer	On the course over Sagami Bay	Retrieval of part of the tail; picture of Yumi's Ochiai's face
Yomiuri Shimbun	JAL plane, 50 bodies are identified	Tail (vertical stabilizer) recovered in Sagami Bay	Wreckage at crash site with SDF soldiers in the background

Table by the author.

the crash site. The *Mainichi's* map is the smallest scale so that it can also include JL123's intended destination of Ōsaka within the map and, to fit with its headline, shows the crash site as being in Nagano. The *Nikkei* and *Sankei* use remarkably similar maps which, despite giving preference to Nagano in their side headline, indicates the crash site to be just in Gunma.

This chapter has discussed how the newspapers attempted to include the passenger and crew list in their newspapers. In the case of the *Asahi*, the passenger list starts on page two and continues on to page three. It has about 62 per cent of the detailed passenger list done, with most of the remaining 38 per cent listed in *katakana* separately. It also has all of the crew listed with facial photographs, presumably provided by JAL. *Mainichi's* passenger and crew list is on page eight and includes facial pictures for 36, with pictures of the crew being included in an article on page 18. Unlike the *Asahi*, the *Mainichi* does not separate out those for whom additional information has not been found (about 55 per cent). Also, unlike the *Asahi*, the *Mainichi*'s list of foreign passengers provides the names in romanized letters rather than *katakana*. The *Yomiuri's* passenger list had details for only about 33 per cent of passengers, but contained the most pictures of passengers (89). The *Sankei* also managed to get 55 passengers photos and had details for about 33 per cent of the passengers. Being a business-focused paper, it is perhaps not surprising that the *Nikkei* seemingly put no effort into getting details of the passengers. Indeed, it only had very brief details for just one businessman and only had *kanji* for 19 per cent of the Japanese passengers, with the rest of the names listed in *katakana*. Whereas the other papers continued to update the passenger list over the following days, the *Nikkei* did not.

Let us now consider the front page of the newspapers on 14 August (see Table 7.2). Perhaps, as evening editions on the 13 August had provided headlines about the rescue of the four survivors and the television news of that day would have covered the rescue in such detail, the newspapers did not particularly focus on the rescue of the survivors. None of the papers featured it as their main story; two allowed the photographs to deal with the story, one had the story of the survivors on the front page with a file photo of Yumi Ochiai, but the other two did not feature the story at all on its front page. As the recovery of part of the tail happened late in the evening, after the main evening television news, the newspapers may have considered this as being their best opportunity to provide a story which readers might have not been aware of when they went to bed on the night of 13 August. Although the *Asahi* has pictures of all four survivors, three of the pictures were taken prior to the crash rather than pictures taken at the crash site and so were provided by a third party, for example JAL in the case of Yumi Ochiai. If we look inside the papers on the 14 August we see that all have a number of pictures (not including pictures of faces for the crew and passenger list, for example, and other file pictures, such as that of JA8119 after its accident in Itami), with between 25 per cent and 50 per cent of them of the survivors and the rescue effort. The largest number of pictures was 11 in the *Mainichi*, whilst the *Asahi* had ten, the *Yomiuri* had nine, the *Sankei* had eight and the *Nikkei* just four. Three of the *Nikkei*'s pictures appeared on one page and took up about 20 per cent of the page, whereas the other papers contained pages with about 33 per cent pictures, although in the case of *Asahi* and *Yomiuri* it was just one large picture of Keiko Kawakami being winched away that took up the space. The relatively high use of pictures demonstrates the scale of the story and the very visual nature of it.

Just three days after the crash there is a big difference between the newspapers' reporting. The *Nikkei*'s morning edition on 16 August dropped JL123 from its main story, turning instead to a business-related story, but kept a front page story about JL123. The *Yomiuri* also dropped JL123 from its main story, focusing instead on a possible increase in defence spending, but kept a front page story about JL123 (about concerns of metal fatigue in the tail). The *Asahi*'s main story is still about JL123, saying that the focus was on the right side of the plane as there appeared to have be something wrong with the plane. The *Sankei* led with the story that there were more survivors and, had the SAR been faster, more people would have lived. The story was based on interviews with the doctors in Fujioka conducting the identification process. Clearly this story was based on investigative work by the journalists and not merely relying on press releases, as otherwise such a shocking headline would almost have certainly have featured in all the papers. However, the *Mainichi* had another surprising headline – that there was first an explosion of the rear bulkhead, and the story suggests that this subsequently damaged the tail of the plane. It also questions whether there could be a link to the 1978 Itami accident. The story is based on comments from the AAIC and also Gunma police's investigation team. Given the discussion in Chapter 4 about the cause, this article is shocking as it is so similar to the conclusions which the AAIC would eventually come to, and more particularly that this was being suggested before the American team had visited the crash site. This clearly contradicts the way that events are presented in many documentaries about the crash. That only the *Mainichi* had this story suggests that they were either able to carry out a significant degree of investigative journalism that the other papers were not, or that they had somehow benefitted from an exclusive divulgence of information.

This chapter has already shown that there was a diversity in reporting of JL123. However, I assumed that one area where there would be greater consistency was in the reporting about the identification of remains. As journalists were not allowed in the morgues, I assumed that they would be reliant on press releases for getting the information, which could lead to similar reporting. Looking at the newspapers from 20 August, we discover that they all contain a list of those recently identified. However, the way they are presented and the order the names appear in are not the same. The *Nikkei*, as may be expected, merely lists the names and ages taking the least space to list the 33 names. The *Asahi* list also includes the city in which they were resident, but the names are not in the same order. The order of the *Mainichi* list is different again and also includes the full address of the victim and their profession. The *Mainichi* is also the only one to note the nationality of the Japan-resident-foreigner identified. The *Sankei* also gives the full addresses of the victims, together with the details of the company where they were employed. The order of the list is different again and only has 28 victims. The *Yomiuri* lists 32 as having been identified and provides their age, city and prefecture where they

Tadashi Sera 瀬良直司 *(36)* · *Hiroshi Shibata* 柴田宏 *(54)* · *Kisaku Shiga* 滋賀喜作 *(45)*

were resident. So, even with a story which was almost certainly reliant on an initial press release of the names identified, differences develop due to the research the newspaper has been able to do about each victim. That differences in ages for certain victims are found is perhaps an indication of how variable the results of this work were.

We see some significant differences if we also look at when the newspapers stopped running JL123 stories on their front page. Unsurprisingly, the *Nikkei* was the first to stop having any JL123 stories on its front page on 21 August, and dropped JL123 as its main story on 15 August and not run it again as its main story after that. The next to drop JL123 from the front page was the *Mainichi*, which first dropped JL123 as its main story on 21 August and again did not have it as its main story on 23 August, but it was not until 26 August that there was no JL123 story on its front page. The *Asahi* first did not have a JL123 on its front page on 27 August, having dropped it first as it main story on 22 August and then again on four more occasions. As mentioned above, the *Yomiuri* dropped JL123 as its main story on 16 August, and it did not feature as the main story on five more occasions before the story was dropped from the front page on 29 August. With its headquarters in Ōsaka, it is not surprising to find that the *Sankei* had the latest date, 30 August, on which no JL123 story featured on the front page, although by this time it had been dropped as the main story on six previous occasions with the first being on 21 August. Naturally, when a new significant story arose, JL123 returned to the front page at a later date outside the period I am focusing on here.

So far this section has concentrated on the morning editions of the five main national newspapers, but there are other papers. First, many newspapers, including all five main national newspapers, have an evening edition. The number of pages is usually less than the morning edition and, as can be seen from Table 7.1, the sales are not as high as the morning edition. In looking at the coverage of JL123 in August 1985, three things are particularly noticeable. First, in order to get the full picture, one would need to get both the morning and evening editions of papers as the morning edition of the next day does not necessarily contain information or articles that were contained in the evening edition on the previous day. Second, some of the articles in the evening editions appear to be less conservative than the morning editions, for example with more articles critical of the SAR. Third, the evening editions are much more graphically based with a higher use of photographs (in some cases nearly a whole page is dedicated to one or more photographs) and graphs and other figures.

In addition to the five main national newspapers there are also numerous local newspapers across Japan. My hypothesis was that these papers would have significantly different coverage of JL123 to the national newspapers. As they have fewer resources and a greater local focus, I expected to see that the coverage of JL123 would be both less than the national newspapers and that articles would be more similar as they became more reliant on news agency reports. The main exception I expected would be *Jōmō Shimbun*, which covers the area where the crash occurred, the area where the *izoku* were waiting to identify loved ones, and a local victim. Due to these factors I would expect the *Jōmō Shimbun*'s coverage to be quite detailed and to have a high number of photographs in order to compete with the national newspapers. However, in doing the analysis I found that in reality it was difficult to find a local paper which was not able to localize the crash. For example, the *San'yō Shimbun*, based in Okayama, contains a number of articles relating to local victims, although officially there were no victims from the prefecture. For example, the *San'yō Shimbun* reports on the Yoshizaki family as being from the area, which leads to greater focus on the two survivors of this family, whilst most other papers focused upon Keiko Kawakami, who was often merely referred to as Keiko-*chan*.[f]

If we consider the headlines of 13 August, we have already noted that the *Jōmō*'s headline is different to that of the national newspapers as it does not give the number of passengers, adds that the plane was a jumbo jet and notes that there was a fire. This last point was perhaps needed as it did not use a picture of the crash site, unlike the national newspapers. As mentioned above there was also mention of the local victim on the plane. Many other local newspapers' front pages much more closely resembled those of the national newspapers, often using the same SDF picture of the crash site, which is indicative of their need to rely other agencies for national news. Although *Jōmō Shimbun* gives preference to Nagano over Gunma as the location of the crash site, its accompanying map of the front page is a larger scale than those in the national newspapers, as the local names would be familiar to readers. Inside there is noticeable absence of passenger and crew lists compared with the national newspapers. Clearly, local newspapers would just not have had the resources for this on that first night, and so such lists appear the following day.

If we look at the local newspapers' morning editions on 14 August, the focus is on the fact that survivors were found. Having a local photographer on the scene also meant that *Jōmō Shimbun* was able to include a picture, which accounts for about half of the page, to go with the story. Indeed, *Jōmō Shimbun* solely reports on JL123 for the first five pages and has another two pages primarily focused on JL123 later in the newspaper. Other than the headlines, pages four and five only include photographs and a map. The largest of these pictures, showing the search through wreckage in Sugeno-sawa, accounts for about three-quarters of the page. On 24 August, *Jōmō Shimbun* further demonstrates its photographic advantage by having a two-page spread of pictures, with one picture representing each day in the 'two-week' period since the crash.

We see a variety of stories being covered on 16 August. Some local newspapers focus upon Prime Minister Nakasone's visit to Yasukuni Shrine on 15 August, whilst many discuss JL123 and the idea that there may have been a problem with the tail of the plane. *Jōmō Shimbun*, on the other hand, features a picture of an *izoku* crying on a coffin and a headline that the analysis on the CVR will begin soon. Looking at the articles in the newspapers on 20 August about the identification of victims, one sees a variety of lists, as was seen with the national newspapers. Looking specifically at the one in the *Jōmō Shimbun*, one finds that they have listed the victim's name, age, address and place of work. However, in relation to the identification of victims, the local newspapers tend to

Takuo Shimada 島田拓夫 *(47)* · *Kinai Shimamoto* 島本喜内 *(61)* · *Yoshio Shimano* 島野佳男 *(41)*

be particularly concerned with the identification of local victims. Indeed, this was the main headline on 19 August for *Jōmō Shimbun*. Additionally, local newspapers appear to be more concerned than the national newspapers in reporting about the identification and the wakes and funerals of the famous victims, such as Kyū Sakamoto. Because of the greater local focus of newspapers, we also find that there is seemingly greater preparedness to drop JL123 from being the top story, which happened on 16 August in some cases, and from the front page altogether, although this did not often happen until around 27 August. The *Jōmō Shimbun* first drops JL123 as its main story on 17 August when it focuses upon the complete opening of the Kan'etsu Expressway, but JL123 is again often the top story on subsequent days and remains on the front page until 31 August, which is later than any of the national newspapers.

Before looking at the media coverage of the Osutaka Pilgrimage, there is one further issue that is worth highlighting. In the wake of the crash, there were many articles which were heavily critical of JAL. van Wolferen suggests that the media 'developed an elaborate campaign in which JAL was portrayed as the great villain' and that 'Not a day passed without the newspapers finding some fault with the national airline'.[41] This was apparently due to comments by Transport Minister Yamashita, who had flown on JA8119 on its flight from Fukuoka to Haneda before becoming JL123 and bought three plastic model planes from one of the stewardesses, who stayed on the plane for the JL123 flight, as presents for his grandchildren,[42] when he indicated that he thought JAL was to blame. If this were really the case, it would be indicative of the close relationship that is thought to exist between the media and the state. van Wolferen goes on to suggest that 'Questions about the disastrous tardiness and manifest lack of effective interagency communication were submerged' amongst the articles attacking JAL.[43] The particular focus of the attacks was not just JAL itself, but particularly its president, Yasumoto Takagi. Takagi was the first JAL president to be appointed having risen up from within the company rather than be appointed by the Ministry of Transport. Consequently he may not have been not popular with the Ministry of Transport,[44] and the media attacks could have been a reflection of the bureaucracy's, and therefore the state's, feelings towards Takagi. Given what I have presented in this section, and based on my studies of the newspapers, I would suggest that van Wolferen's comments are a little exaggerated, but it is certainly the case that the Japanese media perhaps did not question the state as much as it could have done, or should have done.

So why did JAL not defend itself? And how were such articles possible given that journalists should have feared repercussions from the *kisha kurabu*? An answer to the first question from one of those who has worked in public relations in JAL is that 'In Japanese terms we were *kagaisha*, the perpetrators of a crime, and we should behave in a humble posture'.[45] In other words, it would have been inappropriate for JAL to make any defence, regardless of how 'inaccurate and wrong' the reporting was. As for the *kisha kurabu* – this was not a significant concern as the articles were being written by people in other parts of the newspaper companies and not those in the parts that ordinarily dealt with JAL. This provided the additional complication that JAL public relation officials often had to explain technical issues to these journalists as they were unfamiliar with aviation terms because of their lack of experience in this area.[46]

Media coverage of the Osutaka Pilgrimage

Chapter 6 covered what I refer to as the Osutaka Pilgrimage. The media presence at the anniversary events each year is very significant and appears to be disproportionately high compared with some other transportation disasters. This may be in part due to the cocktail of issues that make JL123 of interest, as will be discussed in Chapter 8. However, the relative lack of other political and business-related news at this time of year could also be a contributing factor to what Satō refers to as 'August journalism', with its emphasis on memorialization.[47]

As well as being in Ueno-mura to collect stories, the media help to perform other roles. At the *tōrōnagashi*, lights for the TV camera crews create lighting, provided through power from their own mobile generators, so that people can see what they are doing. Without this it would require the 8/12 Renrakukai or Ueno-mura to provide lighting, which would have a cost implication. Similarly the majority of the lighting at the ceremonies at Irei-no-Sono is provided by the media.

By interviewing the *izoku*, I believe that the media may also be fulfilling another role. The *izoku* did not receive any counselling after the disaster and this has been a role largely performed by 8/12 Renrakukai members keeping in touch with each other through the newsletter or through direct contact with other members (see Chapter 6). Talking to the media is another means by which the *izoku* can talk about their feelings and reminisce about their loved one(s). In this way I believe that the media can help with the continued healing process.

By continuing to report about JL123, it also aids with the aims of many of the *izoku* and the 8/12 Renrakukai of ensuring the accident is not forgotten. This is an important point in terms of the legacy of those who have been lost but also in relation to the aim to promote aviation safety further. However, I am concerned that some of this is done at a rather superficial level. The articles may refer to the tragedy of 1985 and discuss the pain that families have felt as a consequence and these articles may lead readers to considering whether aviation is safe or not, and what can be done to make aviation safer. However, the message is very much an implied one. There is a lack of articles which show that the newspaper has done additional investigations or analysis about aviation safety or the way in which accidents in Japan would be investigated, which has also been a concern of the 8/12 Renrakukai (see Chapter 9), or whether there should even be a reinvestigation into the cause of the JL123 crash itself. This is despite the fact that I have even heard the media ask key people, such as Transport Minister Maehara, when he became the first Transport Minister to visit the site in 2010 (see Figure 6.6k and Chapter 9), about these issues. Maehara himself questioned what such an investigation could achieve after so many years.

There are other things I have observed which mean that not everything the media does can be seen in a positive light. For example, in order to get a good photograph of *izoku* putting lanterns into the river, many photographers will take up spots in the river itself. I have observed many lanterns then hitting these photographers and sinking. Perhaps the most shocking incident I observed was when TBS decided to start a live transmission from Irei-no-Sono at the precise moment that the minute's silence was being held to mark the exact time of the crash. Although TBS thought they had placed the announcer in a location which would not cause any disturbance,[48] the sudden additional light thrown onto the reporter as well as the sound of his voice seemed to be totally disrespectful to the situation. When one considers that it would have been possible to record his words in advance and pair this with the sight of people marking the silence, it also seemed most unnecessary. As Walter points out,

> Producers love anniversaries – they can research and film well in advance, get footage both of visually dramatic ceremonies and of personal reminiscences, and present it on the day as topical news – something very useful for the news team to have 'in the bag'.[49]

Hopefully the media covering a minute's silence will consider using items 'in the bag' in the future.

Watching the media at work is very interesting as they hover around like flies looking for some food. Each year they appear to be looking for a new angle or story to add some variation to the more standard *izoku* accounts of their memories of their loved one(s), what happened in 1985 and what they have done since. As Reader notes, whilst pilgrimage can be a 'social phenomenon', 'it also has a very individual nature' and 'every pilgrim has a story to tell'.[50] Even my own presence becomes an angle of interest for the media and the way in which the media interviews me can be as instructive as my observations of them interviewing *izoku*.[51] There are cases when journalists will move in closer as an interview is being conducted and take notes about some of what is being said. Once the first interviewer has finished, the next will move in to clarify points from the first interview and also often ask supplementary questions, sometimes apparently looking for a slightly different angle for their story. It is also not uncommon to do the same interview many times. With different journalists often covering the anniversary each year, it is not uncommon to end up giving the same interview each year, which must become tedious for some *izoku*. I have seen cases, particularly at Osutaka-no-One, when *izoku* have made it clear that they will not talk to the press. This may mean that other *izoku* tend to get interviewed more frequently as a result. There also appear to be certain families, with the Miyajima family being the most obvious, which the media particularly likes to follow and interview. Although, on balance, I would suggest that the media do behave respectfully towards the *izoku* now, this apparently has not always been the case. Many *izoku* I have spoken to mentioned how in the past they felt harassed by the media and that their questions to children in particular tended to be inappropriate and led to further anxiety for the child and the child's parent. The father of one *izoku* was himself a journalist, and although he apparently felt that they were conducting themselves as he would have done, he was shocked to see what it was like being on the receiving end.[52]

One thing which is noticeable about a number of the interviews is the level of co-operation between the media. This is particularly evident with the interviewing of *izoku*. Perhaps out of a desire not to disturb the *izoku* any more than necessary, it is not uncommon to see one journalist interview the *izoku* and then turn round after the interview has finished and brief journalists from competitor organizations about what was said. This particularly happens at Osutaka-no-One, perhaps due to the limited space at the site. However, it should also be noted that in their need to find a story of interest many more *izoku* are interviewed whose story does not appear in print.

On 12 August, the media are among the first to arrive at Osutaka-no-One. They effectively set up camp and prepare their computers and, those who have them, also set up their satellite-link technology for filing the stories. Most companies have between two and three reporters. One of these might be responsible for taking photographers while the others are responsible for interviewing. Those doing the interviewing in particular might walk between *Shōkon-no-Hi* and the car park several times during the day as they look for people to interview. They could also go with *izoku* to their loved one's *bohyō* or just walk around the site to try to find someone to interview. Although, for much of the day it is *izoku* visiting Osutaka-no-One, later in the afternoon others visit to pay their respects. In the late afternoon the President of JAL will come and lay some flowers and then make a brief statement, which may be followed up with additional information from representatives from the public relations department at JAL. I have also observed eminent professor Kunio Yanagida and the Governor of Gunma coming to the site to pay their respects and holding some interviews. Transport minister Maehara's visit in 2010 is covered earlier in this chapter and in Chapter 9.

In 2007 another angle providing interest for the media was the first attendance at the anniversary events by Mitsuko Fujizaki of the '4·25 *Nettowāku*' ('4/25 Network'), the association for *izoku* of those who died in the Fukuchiyama Line derailment. Whilst, for the media, this provides a means to write fresh, new stories compared to previous years, it also further highlights how JL123 and the Osutaka Pilgrimage has come to be more than simply about JL123 itself. Indeed, such stories may further aid the aims of the 8/12 Renrakukai by ensuring that JL123 remains relevant today and is not forgotten and over-shadowed by more recent tragedies. The JL123 story has become so enduring that it has the power to unite and ensure that the authorities cannot ignore it or other safety campaigns and *izoku* groups. The media has an important role to play in this respect. However, as suggested above I fear that sometimes the media coverage does not go as far as it could do.

Figure 7.7 shows how the number of articles, and their length, covering the anniversary events in Ueno-mura in the *Asahi Shimbun's* morning and evening editions has varied over the years. Compared with Figure 6.10, one can see some similarities in the pattern. For example, the number of articles appears to increase with particular anniversaries (the tenth, the twentieth and particularly the twenty-fifth), but the significance of the main Buddhist anniversaries is less apparent. However, there are other media-specific reasons for some of the results.

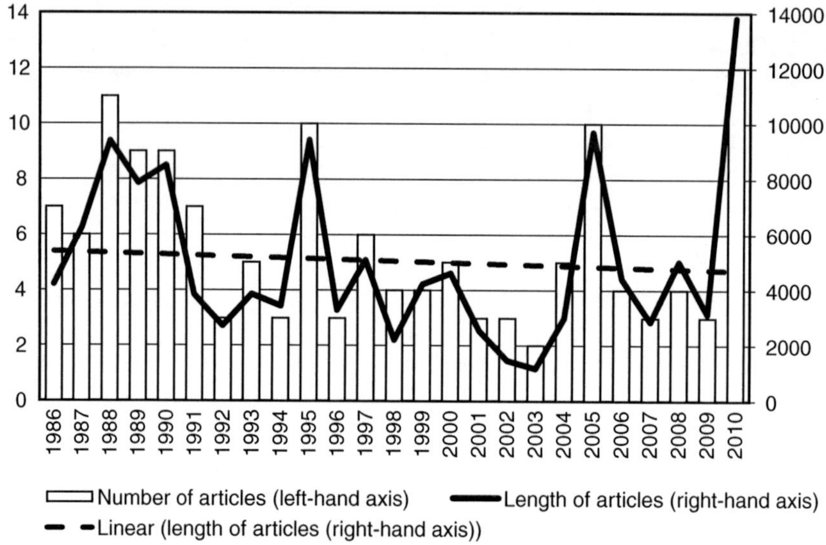

Figure 7.7 Media coverage of the Osutaka Pilgrimage

Figure by the author.

For example, the drop in 2007 was caused by 13 August being a press holiday, so there was no morning edition of the *Asahi Shimbun.*

In comparison to other disasters, whilst media coverage of the Great Hanshin Earthquake is likely to be greater than that given to JL123, it is not unusual for there to be no articles at all for the coverage of the China Airlines flight CI140 crash anniversary, for example. It would appear that there is a difference between short-term impact and long-term impact for certain incidents. The more significant the initial incident, the more likely it is that the media will continue to report it. But what defines an incident as being significant? Rather than merely loss of life, it would seem that other factors seem to be involved. Much of this relates to why JL123, for example, is of interest to people (see Chapter 8), but it also relates to whether there are broader implications for society. In the case of the Great Hanshin Earthquake the earthquake threat to other areas of Japan among other factors is significant, whilst in the case of the Fukuchiyama Line derailment, the press questioned Japan's obsession with punctual train services.

Today, compared with the reporting in 1985, one does not find the same degree of criticism of JAL in relation to JL123. Of course this is in part due to the fact that the anniversary tends to be focused much more on the *izoku* than on any issues relating to the cause of the crash or issues related to that. However, this does not mean that the articles are totally without any bias. For, whilst articles will include reference to it being a crash of a JAL plane, it is unusual to find any reference to it being the crash of a Boeing plane, although articles may mention that the plane was a 'jumbo'. By comparison, where an accident has involved

Kenichirō Shiraishi 白石憲市郎 *(37) · Tatsuya Shirasaka* 白坂達也 *(40)*

an Airbus plane, there has been a greater tendency to mention the manufacturer's name. This is a point which has not been lost on Airbus Japan's CEO, Glen Fukushima.[53] What the reasons are for this bias is beyond the scope of this study, but given that Boeing itself admitted that it was at fault in the case of JL123, it is certainly surprising that the Japanese media appears to be prepared to aid in the protection of Boeing's image compared with Airbus's.

Conclusions

When I began this research I had a relatively negative opinion of the media and journalists. Having become jaded by the level of opinionated articles found in the British media and by the apparent lack of investigative journalism in the Japanese media, I was expecting to see journalists in Ueno-mura behaving without any great scruples or morals. However, my feeling about the Japanese media is much more positive now. The journalists really seem to care about the *izoku* and take a genuine interest in what is said. But as has been discussed elsewhere, there may still be a difference between the individual and the organization. There is perhaps still more that the media could be doing in the way they report the anniversary and to reflect not just what interviewees say but to discuss and lead a debate about the implications of what is said.

Whilst there is a perception of their being uniformity across the Japanese media, this chapter has shown that there is actually a significant degree of diversity. Although one may not find articles with the journalist's opinions included, as can be found in the British media for example, the opinion is inferred by whether the story is included or not, where in the paper the article appears and what the focus of the article is. However, it does appear to be the case that we cannot find the levels of investigative journalism in Japan that one may expect to find in British newspapers, for example. The use of *kisha kurabu* and reliance on high subscription rates may well account for this. However, the case of JL123 also reveals that the media are capable of doing investigative journalism if the conditions are right, but that there are still some limits to what will be written.

Another thing which cannot be forgotten is that many journalists travelled to Osutaka-no-One in the wake of the crash and so were exposed to the same terrible sights that the SAR teams were. Many spent long hours and days at the site, and some even slept there.[54] Many have consequently suffered PTSD. I have come across one journalist who can remember what he was doing on 12 August 1985, but has no memory of what he did for the next two weeks. How one journalist appears to have addressed his own experiences of August 1985 at Osutaka-no-One will be discussed in Chapter 8.

Whilst this chapter has focused primarily on the newspaper reporting of JL123, it is important to remember the significance of the TV reporting also. Whether in 1985 or for the anniversary, TV reporting has been as ever present. Indeed, the immediacy of their images may have helped define the tragedy. These images have influenced films, featured in documentaries about the crash and even in YouTube videos. These are some of the areas to which the next chapter turns.

8 Japan's *Titanic*

> History is the memory of a nation
>
> (J.F. Kennedy)

So far we have looked at the events that took place in 1985, the remembrance events in subsequent years, and the way in which the media has reported issues relating to JL123. This chapter will focus on the more general interest that exists in JL123, including the interest in JL123 outside of Japan.

When I sent the proposal for this book to the publisher I was asked for a single sentence which would explain it. My answer was that 'JL123 is Japan's and the aviation world's equivalent to the *Titanic*'. This chapter will further help to set out why I believe this is the case. So far we have looked at some of the similarities, for example the scale of the disaster and the lack of agreement about the cause, but this chapter will highlight some of the other reasons for the analogy. The story of the *Titanic* is associated with 'conspiracy theories', coincidences and blockbuster films. Indeed, it is probably the existence of these elements that have cemented the *Titanic* in the minds of people around the world. The *Titanic* is so much a part of 'social memory' in many societies that most people probably have no memory of when they first heard the story, just as they cannot remember when they first heard stories such as *Sleeping Beauty* and *Cinderella*. This chapter will look at these issues in relation to JL123, for I believe it is becoming a part of Japan's 'social memory'. It is a story which more and more Japanese, if they are not of an age that they can remember the event or were not even alive when it happened, are becoming familiar with. JL123 may not have reached *Titanic*'s level yet, but this chapter argues that the seeds have been sown for this to happen. To aid in considering these issues, the chapter begins by discussing what is meant by 'history' and 'social memory'. The chapter also considers how JL123 may be remembered by future generations.

History and social memory

What do we mean by history? I studied history at lower secondary school, but all this covered was World War I and World War II. There was nothing prior to

Hisanobu Suekawa 末川久信 *(59)* · *Kazurō Sugai* 須貝和郎 *(38)* · *Akiko Suita* 吹田暁子 *(46)*

the twentieth century, nor was there anything post Hiroshima and Nagasaki. We did not learn about how the British Empire was created, nor how it fell. Indeed, we largely only studied British victories, and there was no sense of discussion about what made the British Empire 'good', as was implied, but just why the German Empire was 'evil'. There was no place for the Falklands/Malvinas War, which had taken place only a few years previously, let alone conflicts that did not directly involve Britain (e.g. the Vietnam War), the creation of NATO and the development of the Cold War, or non-combatant issues (e.g. the creation of the European Community, as it was then). For those countries which were involved in World War II, 1945 has seemingly become a barrier.

But time moves on. Many first-year undergraduate students reading this book in its first year of print would have been only about eight years old when 9/11 occurred. Most have no memory of it, and, at the time, they would not have been able to contextualize it and understand it in light of what had taken place in the years beforehand or what the world's response to it was over the following few years. Many events which seem critical to our own understanding of the world and our experiences are mysteries to others. The same can be true of books we have read or films we have seen. It is essential to keep this in mind when teaching history in any form so that the links can be made more effectively and to ensure that the significance of what is being discussed can be understood. Let me give an example of the challenge. In 2007, the *isho* written on JL123 could have been discussed in relation to the mobile phone calls made from United flight UA93 on 9/11. At that time many more students would have remembered 9/11, and the story of this flight had just been released as a film, *United 93* (2006). In 2010, however, I find that most students have not seen this film and have little awareness of what took place on 9/11's 'forgotten flight'.

Despite 9/11 being 'history' for them, many of my students, when asked to discuss 'recent changes' in the Japanese education system, will include discussion of the occupation reforms in 1940s and the reforms initiated by Nakasone in the 1980s. Clearly, 'recent' and 'history' have variable definitions depending on the context. Yet, if we consider the experiences of current students and what, to them, might seem like a 'long time ago', at the very least it seems it is time to do away with the 1945 barrier and find a more recent cut-off date.

The teaching and study of history should not be merely a list of names, dates and events; they need to be understood in context. Indeed, this goes back to the original meaning of the word. It was not until the nineteenth century that history became 'a social science and sharply differentiate itself from story-telling' and its original meaning of 'A relation of incidents'.[1] A good story and good history should be able to contextualize the information so that the reader or listener can understand the situation. For example, it is easy to be a pacifist when surrounded by peace and to condemn the bombing of Hiroshima from the distance of 65 years, but what if it were your father, your brother or yourself that was about to be asked to be part of a land invasion of the main islands of Japan, with the knowledge of how difficult the battle for Okinawa had been? Hindsight may be a wonderful thing, but it has its limitations if it is not fully informed, with the information being internalized and felt. Perceptions of reality vary with experience and time. In relation to JL123 it is easy now to criticize the SAR activities, and, although I fundamentally believe that a better response should have been possible, one has to appreciate the problems that could be caused by the sheer scale of this unprecedented accident.

The presence or lack of a personal connection to a historical event can be significant. A lack of connection can make it seem remote and hard to understand, but there is also a danger that a connection can lead to an inability to comprehend the event objectively. This is a challenge I have had with conducting this research. The *izoku* and many others that I have spoken to have an obvious connection to the crash. Their memories of events in 1985 have, in some cases, I believe, come to be shaped by experiences and knowledge gained in subsequent years. In the time that has passed there are also cases where things cannot be remembered or where memories become confused. Consequently, one has to be careful in how much reliance is placed on interviews or other materials which rely on interviews. It is not always possible to interview enough people to seek a probable sequence of events based on accounts which corroborate each other. Such corroboration may even be a result of memories being overly influenced by reading other accounts. Although there is a tendency to trust books, these are not without their problems as will be discussed below.

However we define and understand the term ‘history’, what is certain is that some events go on to become engrained within what can be described as ‘social memory’. Vinitzky-Seroussi points out there are problems of terminology when discussing memory as authors variously use terms such as ‘collective memory’, ‘popular memory’, ‘cultural memory’, ‘public memory’, ‘collective representations’, ‘shared memory’, ‘communal memory’ and ‘social memory’, whilst meaning largely the same thing.[2] But this social memory may not be accurate for there is a difference between history and memory as Halbwachs,[3] for example, has pointed out. Social memory may become such a powerful force that it can even influence attempts to write a correct history. For example, the builders and owners of the *Titanic* never claimed that the ship was unsinkable,[4] but this is not what we learn today. Berlinguez-Kōno also suggests that the social memory relating to Takamori Saigō has hindered ‘the scientific writing of history’ and the ‘official narrative cannot be easily modified, as each intermediary group contributes to the formation of a collective memory that often has a political interest in creating “their” memory’.[5] As Young points out,

> Remembrance as a vital human activity shapes our links to the past, and the ways we remember define us in the present. As individuals and societies, we need the past to construct and to anchor our identities and to nurture a vision of the future… However, we know how slippery and unreliable personal memory can be, always affected by forgetting and denial, repression and trauma, and, more often than not, serving the need to rationalize and to maintain power. But a society’s collective memory is no less contingent, no less unstable, its shape by no means permanent and always subject to subtle and not so subtle reconstruction.[6]

Vinitzky-Seroussi suggests that 'Commemorative narratives – particularly the ones of painful pasts – can be understood as consisting of three components: the protagonist(s) being commemorated, the event itself, and the event's context'.[7] In this respect, part of JL123's significance would appear to come from the normality of those who were on board. It is through their death and the way in which they died that they have become extraordinary. The accident came in a year when Japan was already reflecting upon its past, and what Maier has called a 'hunger for memory' was developing.[8] With the moves to the urban areas that had taken place during the previous decades, the *furusato* boom was a reflection of this desire to remember an idealized view of the past.[9]

Assmann talks about the importance of external media and institutions in the forming of cultural memory and includes within this such things as texts, pictures, buildings, landscapes, festivals and rituals.[10] Saaler and Schwentker point out that 'while the Japanese experience is only rarely considered in European research on memory, Japan is certainly one of the "major players" in what Pierre Nora has called "the age of commemoration", with its "ardent, embattled, almost fetishistic 'memorialism'"'.[11] For example, in Japan, how to remember the bombings of Hiroshima has been an on-going debate. The changes inside the main museum have reflected the various changes in attitudes and the conflicts among the 'community of memory', i.e. the *hibakusha* (atomic bomb victims) and the conflicts with the nation.[12] Although the remembering of Hiroshima has become a national, if not international, project, much of the driving force has come from the local level.[13] In the case of JL123, there are differences among the *izoku*, as discussed in Chapter 6, but the main conflict has been with JAL rather than the state, which has largely deferred the matter to the *izoku* and JAL. Indeed, it was not until the twenty-fifth Anniversary that a Minister of Transport went to Osutaka-no-One on 12 August.

So why do some events get remembered, whilst others are consigned to history and forgotten? The *Titanic* is well known and commemorated with a plethora of memorials; five for crew in Southampton alone. However, the *Princess Sophia* which sank in the inside passage route between Vancouver and Skagway, with the death of all 353 on board, has no memorials other than tombstones where bodies were recovered.[14] This chapter will look at some of the possible reasons why JL123 is engrained, and perhaps is becoming more so, within the social memory of Japan.

Conspiracies, coincidences and the butterfly effect

In relation to JL123 there are a number of alternative theories about the cause of the crash. Many of these could come under the umbrella term of 'conspiracy theory', but this term has a particular nuance: that there was a deliberate effort, often by the state, to cover up the truth. 'Conspiracy theories' have become a popular genre of writing and a key component of many popular TV programmes and films. The enduring nature of some of these theories may even be the reason why some events, whether it be the assassination of JFK, the sinking of the *Titanic* or the death of Lady Diana, continue to remain of such interest to many.

Mitsuru Tabuchi 田淵満 (19) · Junko Tabuchi 田淵純子 (14) · Noboru Tachibana 立花昇 (45)

In terms of conspiracy theories related to JL123, Masaaki Ikeda is one of those who has put forward the idea that JL123 was in fact shot down in what he, in one of his many books on the subject, claims was 'the perfect crime'.[15] To be 'a perfect crime', Ikeda suggests that there are five necessary conditions: first, the one pulling the strings and those who committed the crime have not been caught; second, the motives are unknown; third, an alibi has been established; fourth, evidence has been destroyed; and fifth, the statute period of fifteen years for murder has passed.[16] Although Ikeda suggests that JL123 was shot down by the SDF,[17] the fact that he suggests that one of the conditions to it being a 'perfect crime' is the lack of certainty about the motive would appear to be just one of the weaknesses with his argument. Ikeda discusses three possible motives; first, for self-satisfactory reasons; second, avert the people's attention away from a crisis in the state's control; third, that it was the first step in a militarism using new, extreme weapons.[18] Ikeda continues to write about the subject, and although he does not want to believe the SDF shot JL123 down, he believes that it is important that the world discovers the truth about what happened.[19] Whilst Ikeda's suggestions seem extreme and hard to accept, there is a question about some materials found at the crash site which were not connected to JL123 and it is suggested that these materials were part of an SDF rocket.[20] The first and the most regular type of emails I received after putting details about my research on the internet was from people in Japan who believe that JL123 was shot down. The comments in one email are typical of the views put forward:

> There is a lot of evidence of crashed pieces of missiles and bodies of military fighters and helicopters have been found... This is not only a domestic crash but also an international military problem... Please note that our team is under critical observation of the Japanese government.[21]

So, is it possible that JL123 passenger Tetsu Ogawa was trying to take a photograph of a missile? The first picture taken from inside the plane was an angle suggesting it was taken by his daughter. However, the other four pictures are taken from a different angle, suggesting that Tetsu Ogawa asked for the camera back. The angle would also suggest that the pictures were taken by someone standing – if it was taken by the person in the seat by the daughter, the angle and contents of the picture would likely have been different. But what was the purpose of these pictures? If the intension was to get a picture of Mount Fuji, which appears in the background, why not allow his daughter to take them? If he was standing, it also means that either the seat belt sign was off or he chose to disobey the sign. Either of these possibilities are significant for it is generally thought that the seat belt signs were never switched off for passengers; whilst to disobey the seat belt sign would suggest that what was being photographed must have been of great significance. Figure 8.1 is one of the pictures taken. When Mr Ogawa's son showed the picture to photographic experts, they increased the resolution and found that the 'mark' below the wing was not a mark on the photograph, which does not appear in any of the other pictures, but an orange object.[22] Orange is the colour of the

SDF test missile, *Fire Bee*.[23] I have also checked the picture at high resolution and the object clearly has some orange colouration to it. But if Tetsu Ogawa saw a missile, surely others did too. Why is there no mention of them in any *isho*? More significantly are we really to think that none of the crew saw such an object, or that if they did, they did not believe that they should report it to ACC? The risk of such a communication would surely have meant that if JL123 were struck by a deliberate missile attack, surely a further attack would have been made.

If JL123 were struck by a missile, perhaps it was accidental and the SDF attempted to cover up the error, concerned about the impact it would have on public perception of the SDF. After all, an SDF error had accounted for Japan's worst aviation disaster to date. Consequently was the delay getting to the crash site deliberate so that the SDF could get to Osutaka-no-One to clear away vital evidence? I find such an idea a little hard to believe given the complexities of the crash site itself and the problems of finding anything among the wreckage, let alone a few key pieces of missile.

A large part of JL123's tail fin that was retrieved pointed to the damage being from the internal forces rather than an external force. This would appear to negate the idea that the plane was struck by a missile, at least on that section of the plane. It also means that the suggestion that the plane were struck by a small meteor,

Figure 8.1 A view from JL123

Photograph courtesy of Ryōichi Ogawa.

related to the Perseids meteor shower which peaks in August each year, is also invalid.[a] However, it would still leave open the possibility that a bomb on board had caused the initial damage. Given the history of terrorism related to air-travel in Japan (see Chapter 2), it is perhaps surprising that so little attention was given to this possibility. Although JAL received a call from a group, Kakumaru, claiming that 'it was our power that blew up the plane',[24] this possibility appears to have been discounted. The AAIC report, in a rare example of discussing something not directly concerned with proving its own conclusions, points out that there was no evidence of any explosive device having been detonated.[25]

By coincidence the retrieval of part of the tail of JL123 in Sagami Bay by the SDF happened almost exactly 24 hours to the minute after the plane crashed onto Osutaka-no-One.[26] This is by no means the only example of coincidences. As mentioned in Chapter 2, a message was received by police on 12 August 1985 that the campaign against companies such as House Foods was ending. Coincidentally House Food's president, Ikuo Urakami, was one of the passengers on JL123. By chance an amateur photographer had videoed JL123's take-off.[27] It was also just by chance that Keiichi Yamazaki had his camera to hand to get the last picture of JL123 (see Figure 4.5). Whilst Yamazaki won the Yomiuri Photograph Award for the picture, it did not come without costs as some people phoned his house to say that it was his taking of the picture which caused the crash.[28] Around the time of the crash itself, apparently Captain Takahama's eldest daughter stumbled and broken a heel of her shoe.[b, 29] In another case, a house suffered a power cut as the person's loved one perished in the crash.[30] Another *izoku* says that, at about the time of the crash, she heard a voice in her head saying that her father was not coming back. When she dismissed this thought with the idea that her father was going to be away until the end of the week, the voice said that 'your father will never return'.[31] Coincidentally, in the in-flight magazine on board JL123 there was an article which included an interview with Takeo Kurosawa, mayor of Ueno-mura, which had been conducted some two months earlier.[32] The interview contained the words which were apparently said to Kurosawa at the start of 1985 by one village elder; 'This year, something will occur which will make you the attention of the whole world'.[33] Following the crash, all the copies of the in-flight magazine were destroyed.[34]

As noted in Chapter 1, it was partly due to my own coincidences in relation to JL123 that motivated me to start the research. After I began the research, I was surprised that in the first Japanese documentary that I watched one of the few interviews in the programme with *izoku* was with the brothers of the *izoku* I had first met more than 13 years earlier and that the programme contained a family picture with this friend in. Once I began telling people about the research I was struck by how many could connect themselves to the crash in some way. Some of these people had taken a JL123 flight before. Others, without being *izoku*, knew people on the flight or know of people who knew people on the flight. I find it intriguing how many people can connect themselves to the flight in this way like a form of 'six degrees of separation'. I also find it interesting how there is sometimes a desire to be able to do so. After three years I began to be more surprised when if, after telling a Japanese person about this research, they could not think of anything which connected them to JL123. The coincidences and connections kept coming. One of the most bizarre, was as I wrote the sentence in the book about the photograph taken of Mr and Mrs Miyajima at the crash site, an email from Yomiuri with that photograph attached for me to use in the book arrived in my inbox. Then, at a dinner after a workshop presentation, I was sat with two Japanese people who were born on 12 August 1985.

But some may suggest that there is more to it than mere coincidences and that the changing of any one event, however minor, could change the ultimate outcome. This idea is often referred to as 'the butterfly effect' and has become popularized through novels and films which look at the impact of changes being made as a result of someone going back in time. Let us consider two examples of this theory in relation to JL123. First, back in 1978 some young men formed a group called Echo and the Bunnymen. In due course, a man called Kimble Mathews would do some work for the group. Kimble then met a Japanese girl, Masako Nishiguchi, at one of the concerts. Kimble got a job in Japan and the couple later became engaged. After returning from a trip to Europe in August 1985, Kimble and Masako transferred to JL123.[35] Would changing any of these events have led to a different conclusion or was there an inevitability of these two being on JL123? Second, what if Nakasone had taken greater responsibility for JL123? What if he had resigned? This could have meant that a different Prime Minister may have represented Japan at the next G5 summit. Would another Prime Minister have signed the Plaza Accord? If not, would the bubble economy and the subsequent recession have happened, or happened when it did?

What about those who just managed to catch the plane, and the others who were unsuccessful in getting tickets, despite there being a few empty seats? Was this just an issue of luck? Interestingly, 'The Butterfly Effect' was also the title of an album released in 2010 by Diana Yukawa, who was born a few weeks after her father died on JL123. The album features a track included in the film *Shizumanu Taiyō* (2009).

Mention was given in Chapter 6 to astrological theories related to JL123's fate and also the significance of 'lucky days'. Some Japanese also believe that there are certain 'unlucky ages' (*yakudoshi), which* for men are 25, 42 and 61, and 19, 33 and 37 for women. However, as with the counting of years for memorialisation after death, the first year is considered to be '1' rather than '0', so the actual age of the person is one less than the *yakudoshi* age. As can be seen from Figure 8.2, the most common age of the male victims was a *yakudoshi* age. However, the other *yakudoshi* ages are not similarly represented among the JL123 victims, indeed there were no female victims aged 18. There were also some other numerical coincidences and issues. For example, nine passengers were seated in a row number corresponding to their age. Subsequent to the crash, as is common practice around the world, JAL retired that particular flight number, although this did not happen straight away and passengers on a JL123 flight on 22 August were featured in a photograph and article on the front page of the evening edition of the *Yomiuri Shimbun* on 24 August. It is not uncommon to come across English

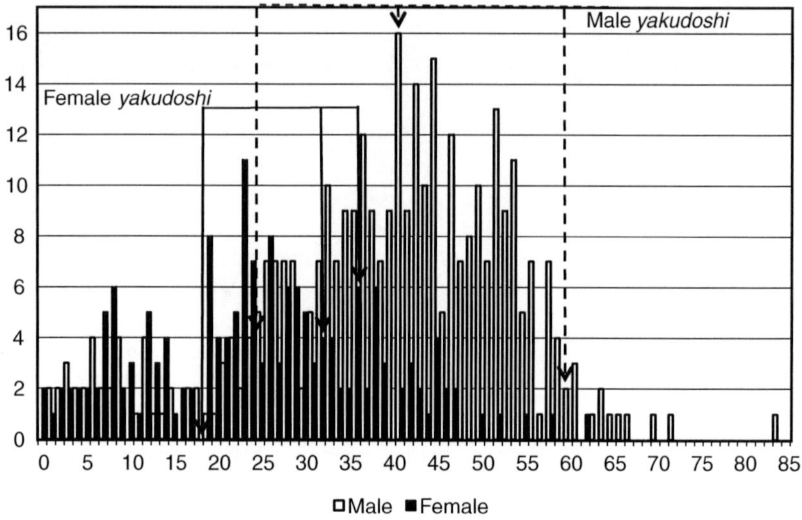

Figure 8.2 Age profile of passengers and crew

Figure by the author. Data from Irei-no-Sono Zaidan 1988.

phrase books for Japanese travellers which use 'Flight ABC 123' as an example. Perhaps the fact that the number of the flight was so simple and easy to remember is one of the factors that has helped to keep the memory of this crash alive. In relation to numbers, it is also worth reflecting on the fact that I have come across *izoku* who, whilst waiting in the gym in Fujioka in 1985, used to look at their watch at 18:56 each day.[36]

Narratives on JL123

As well as the coverage that the media give to the anniversary events, and any other significant stories relating to JL123 throughout the year, books relating to JL123 have been published in Japan since only a few months after the crash. The first, *Ima Hikōki-wa Anzenka*, which suggests a link to the crash through use of photographs of the crash site on its cover, appears to have been written prior to the crash as there is no mention of the crash in the text. In some cases the title suggests that the whole book is about JL123, but in fact it may only be a part of the book. But there is no let up in the number of books relating to JL123 being written. Indeed, if anything, there appears to be an increasing number of books being written, although there have been particular years – notably around the time of the twentieth anniversary – when there has been a particular increase. I suspect that this trend may continue for some time. Some publications have been as a result of policemen, for example, involved with JL123 retiring and recounting their experiences. Such publications are likely to continue whilst there is a perceived market

Nobuyuki Takeshima 竹島伸幸 (25) · Motoaki Takeshita 竹下元章 (47)

and interest in the crash, particularly as many Japanese keep a diary which may contain pertinent information that can be used as a basis for the publications.

If we only consider books which have ISBN numbers and so are most widely available for purchase, there have been over 80 books written either on JL123 exclusively or have significant contents (sections or chapters) relating to JL123. Nearly half of the books are concerned with investigating aviation accidents, with most concluding that the official report's conclusions were mistaken. Of course this in itself may not be surprising since there is probably little need to write a book that confirms the official report, unless it can do so in a way which presents the material in a more comprehensible manner or which has additional evidence to present. About 20 per cent of the books have been written by *izoku*, including titles edited by the 8/12 Renrakukai and books by individual *izoku*. These books can be seen as an extension of the memorial activities described in Chapter 6 for they often focus on their memories of their loved one(s), include calls for ensuring that the crash is not forgotten and that aviation safety is improved, discuss what human moral lessons can be learnt from such a tragedy, and cover with how people cope with the loss of a loved one. To date there has been a relative absence of accounts from JAL employees, other than the novel *Shizumanu Taiyō* (see below). There has also been a lack of accounts by SDF soldiers. Indeed, it was not until 2010 that a book solely concerned with SDF accounts came out. This was also the first *manga* totally about JL123, thus filling another void since this is a popular genre in Japan.[c]

One thing that is noticeable about many of the books is their lack of referencing, bibliography and indexes. Although it is certain that many are based on extensive research, there are times when the texts come across as long opinion papers and those who are interested in knowing how their position was reached are left wondering where some of the information was obtained. While the relative lack of clarification on sources does not mean that such books are not without uses, their reliability is perhaps lessened in the eyes of those who are used to having demonstrate their sources. It is perhaps of little surprise that one of the best referenced books is by an *izoku*, Dr Takao Kawakita, with a more scientific background.

One has to be careful with all of these books. As time passes, there is a danger that mistaken memories can be included in books and presented as fact. For this reason alone, having greater clarity about sources of information would be desirable. Unusually, Kawamura points out some of the errors that others made[37] – although one cannot be certain that his account was less prone to error than the others. However, Kawamura does appear to be correct in pointing out that Iizuka's infamous claim that there was a case of a head found seemingly with three eyes, which was subsequently discovered to be two heads that had been smashed together,[38] did not happen.[39]

Among the books relating to JL123 there have been two novels – *Kuraimāzu Hai* and *Shizumanu Taiyō*. These books are significant for, as Ascari points out, 'it is in novels where society and people are analysed'.[40] The two novels are very different. *Kuraimāzu Hai* follows a journalist, Yūki, as he is put in charge of covering the JL123 crash. However, there is a second story within the book. Yūki was due to go to climb a mountain, Tanigawa-dake, with a friend on 12 August 1985, but the news of the crash broke just as he was about to leave the office. The friend, meanwhile, is hospitalized with a coma after collapsing in the street. He never recovers, and Yūki goes to climb Tanigawa-dake with the friend's son 17 years after the crash. The title of the book is taken from the world of climbing and is based on the English phrase 'climber's high', which is apparently the feeling that climbers get when excitement numbs all fear. The book progresses by flicking between the present and 1985. Although the focus of the book is on the reporting of the crash, the main issue that it highlights is the importance of human life and how, for the media, some lives appear to be more important than others.

The author of *Kuraimāzu Hai*, Hideo Yokoyama, was a journalist at *Jōmō Shimbun* in 1985 and covered the JL123 crash. Because of this it is often assumed that the book is based directly on his experiences and it is seen as being part of the popular *moderu shōsetsu* (fact-based fiction) genre. However, Yokoyama himself stresses that the book is fictional although there are 'pillars of truth' which lead the reader to believe more easily the fictional elements which come between.[41] These 'pillars of truth' are based on not only the facts of what happened in 1985, but also a certain degree of 'universality' based on norms and Yokoyama's experiences as a journalist. Accordingly, although Yūki is not directly based on Yokoyama's experiences, I would suggest that it is possible to see elements of Yokoyama in Yūki's character as well as that of one of the journalists, Sayama, who goes to the crash site. Indeed, Yokoyama says that the article in which Sayama describes his experiences at the crash site (see Figure 8.3), and which is the only piece of supposed newspaper prose contained in the book, is what he wished he could have written in 1985, but it took many years to be able to express it in this way.[42] Although the characters in *Kuraimāzu Hai* are fictional, as it is based on real events, and there is no particular judgement being made about the non-fictional organizations, such as JAL, mentioned in the book, hardly any names need to be changed. In this respect, it is different from *Shizumanu Taiyō*.

Shizumanu Taiyō ('the sun which does not set'[d]) follows the life of one employee, Onchi, at an airline company, National Airlines (NAL). Due to the nature of its contents, the names of virtually all the characters have been changed, although in many cases it is not difficult to work out who the fictional version is based upon. The story is split into three parts and five volumes, with the third volume being the 'Osutaka-yama volume'. Although apparently a piece of fiction – the Osutaka-yama volume makes it very clear what the basis of this part of the story is. The transcript for the communication between ACC and NAL flight 123 is identical to that between ACC and JL123, for example. Furthermore, the names of *izoku* and passengers mentioned in the text have not been changed. Notably the text mentions Ken Miyajima, who seems to have become a means by which to understand the crash, much as Anne Frank's story allows us to 'grasp the awful reality'[43] of what is being discussed. Apparently, when the author, Toyoko Yamasaki, wrote the book a different surname had been used, but the publisher contacted Mrs Miyajima and suggested that it would be a 'shame' if the real name were not recorded in what was likely to become a widely read book, and so Mrs Miyajima consented to the change being made.[44] Compared with *Kuraimāzu Hai, Shizumanu Taiyō* deals much more with JL123 itself – the flight, the crash and the SAR. Yamasaki did a considerable amount of research – including borrowing *Osutaka* newsletters and other materials from the 8/12 Renrakukai – and this gives the text a high degree of believability,[45] although Kawamura, for example, suggests that it contains errors,[46] and there are conversations, for example, which the author almost certainly had to create. Because of its high level of believability, one cannot ignore that such a book may have a significant impact upon people's knowledge and perception of what happened to JL123.

Both novels were very successful, with *Kuraimāzu Hai* coming second in the first Honya Award in 2004.[47] Despite the plethora of other books about JL123, it is probable that the two novels would be the main exposure, outside of media reports about the anniversary, that many Japanese would have about the crash. Their contents, therefore, are significant as they, to some degree, direct the reader's impression of what happened in 1985. Whilst *Kuraimāzu Hai*'s stance in relation to JL123 is relatively neutral – in that it does not focus upon what happened inside the plane or the cause of the crash – *Shizumanu Taiyō*, on the other hand, even mentions the suggestion that the plane may have been shot down by the SDF.[48]

Given the success of the novels, it should be of no surprise that both have been dramatized. However, the way in which this has been done has led to some notable shifts in the possible impact on the social memory of JL123. *Kuraimāzu Hai* was first dramatized by NHK which made a two-part TV movie in 2005. Although it does not contain all chapters of the book, it is very faithful to the original story and became the first NHK drama to feature the name of a real company (JAL).[49] A feature film was subsequently made and released in 2008. This version contains some notable changes from the original story. For example, the significant storyline about the importance of certain lives over others for the media is essentially dropped. Another journalist has their role emphasized further, whilst their gender is also changed to being female in an attempt by the director to make the film more attractive to a contemporary audience.[50] However, another significant change is that the name of one passenger, Ken Miyajima, becomes highlighted in the film. There are a variety of other changes, but in relation to the film's impact on social memory, it is the end of the film which sees the greatest shift from the novel, as text explains that whilst the official investigation found the cause to be due to a faulty repair in the rear bulkhead (as gets discussed in all three versions), there are many who feel that this could not be the correct cause and that are calling for a reinvestigation. The director says that he thought this was necessary having read some books about the crash and what they said in relation it.[51] The text came as a big surprise to the *izoku*, who had not been involved in the production of the film in any way after the initial notification that it was to be made.[52] The text, as well as some other scenes relating to the crash and SAR which are not in the original novel, also caused some issues with Yokoyama, who was concerned that the storyline was focusing too much on the crash.[53]

Whilst the film of *Kuraimāzu Hai* highlighted its link to a real event, the film of *Shizumanu Taiyō* did the opposite. The film itself came out in 2009, although this had not been the first attempt to make it.[54] The story revolves heavily around the management of the airline company and includes discussions about the relationship with its unions and possible corruption within the company. Being obviously based upon JAL naturally meant that production would be difficult. The film covers the whole story and not merely the Osutaka-yama volume. Having cost ¥2.8bn to produce,[55] it had one of the biggest budgets of any Japanese film. The film runs for three hours 22 minutes and includes a ten minute interval. The production company, Kadokawa Pictures, was obviously concerned about any suggestions that the film was based on real events and stressed that it was based on the novel, itself a work of fiction and so there is 'no connection' between JL123 and the film.[56] However, even within the DVD's bonus features there is admission of the link with the real events when on the final day of filming (which was done on a recreation of the crash site near Takasaki) the cast and crew faced the real crash site and held a minute's silence. The DVD also shows that the lead actor, Ken Watanabe, went to Irei-no-Sono to pay his respects. A further link came with the intermission music being played by Diana Yukawa (see Chapter 5), after a direct request made by her to the director to be involved.[57] However, compared with the novel, the film distances itself from the real events by changing the names of the *izoku* and passengers – including changing the name of Ken Miyajima. But if Kadokawa really wanted to distance the film from JL123, it could have done more. For example, the flight number could have been changed, the name of the mountain on which the plane crashed could have been changed, and the location of the morgues could have been altered from Fujioka. That none of these things happened, and that so much is faithful to reality, would suggest that in fact Kadokawa wanted this to become a definitive account of JL123.

It is perhaps surprising that the film versions of the two novels were made, as although the novels were in the public domain, by their very nature, the films have an ability to be more graphic. It is also possible, particularly when global distribution is considered, that their impact could be greater as they are likely to be viewed by more people than would read the original novel. Although JL123 is part of history, it is, in relative terms, a fairly recent event and there are many *izoku*, let alone the four survivors, who are still alive. Although films about the *Titanic* were coming out only a few years after the sinking, when cinemas were still in their infancy, it was not until 1958 that a film, *A Night to Remember*, with more widespread appeal, which successfully put fictional characters into a factual event whilst also pointing out how more could be rescued, was made.[58]

Perhaps in part because of the timing, neither of the films are particularly graphic in their portrayal of death. With *Kuraimāzu Hai*, as the story focuses on the reporting of the accident, there is little need for graphic portrayal of the crash site itself. The main imagery in relation to the death of a passenger is when a dead girl is seen being carried by an SDF soldier while an article by Sayama is read out (see Figure 8.3). Whilst the NHK uses actual footage from 1985 as a backdrop to the scene,[59] the film version recreates the crash site and in this scene

On Mt Osutaka

The young SDF soldier stood rigid holding a little girl tight.
A red dragonfly hairpin.
A blue polka-dot dress.
Her thin light-brown right arm hung down.
The soldier looked at the sky.
How could it be so blue?
How could the white clouds float there?
How could the birds sing, and the breeze waft over the ridge?
The soldier lowered his eyes into hell.
Somewhere was the girl's left arm, which he had to find.

Figure 8.3 The Sayama article

Picture reproduced with permission from Climber's High Film Partners.
Text taken from Yokoyama 2006:102–3.

clearly shows a severed arm dangling from a tree, paralleling one of the magazine images mentioned in Chapter 7. Neither version includes a scene from the novel in which there is discussion of including an image such as those in the magazines, one of which a character is reading. Other than the Sayama article scene, no dead passengers are seen directly – but there are references to them and to body parts. There are also other deaths, for example the film version shows a reporter, apparently unable to overcome trauma brought on by going to the crash site and seeing dead bodies for the first time, being run over. Both versions contain references to what was seen by reporters at the crash site, images of the makeshift morgues in Fujioka – both using still photos in relation to what story to run one day, and in the NHK version it includes images showing coffins being carried to the morgue from actual TV news footage – and one of the Kawaguchi *isho*, with the NHK version showing a copy of the note itself. At the start of the NHK production, a few black and white still images are shown in a brief overview of the crash. Here, and in the images of the crash site in the film, the dead bodies are wrapped in blankets. I believe that, by not being overly graphic, the scene showing the dead girl helps to focus the mind on the scale of the tragedy whilst not detracting from the main story or its themes.

Shizumanu Taiyō, like the film version of *Kuraimāzu Hai*, recreated part of the crash site. However, whilst both included part of one wing, they made subtle alterations from reality. Whilst with JL123 much of the 'J' of 'JAL' was not visible (see Figure 4.3), in both films it is possible to see all three letters ('JAL' in the case of *Kuraimāzu Hai* and 'NAL' in the case of *Shizumanu Taiyō*). Interestingly, in both films one can see Hanshin Tigers memorabilia. As mentioned in Chapter 6, many people began supporting the team after the crash, and so the use of this memorabilia may be a means of providing a further connection to something that some of the audience will recall of 1985. However, the films overlook the fact that the surge in support was subsequent to the crash.

Unlike *Kuraimāzu Hai*, *Shizumanu Taiyō* also recreates the plane using computer generated imagery (CGI) and has a mock-up of some of the inside of the plane. However, the actual moment of impact is not shown. Instead, the crash is entwined with another storyline in which the main character hunts an elephant. As the elephant collapses the sound of 'NAL123' crashing can be heard, and the two jumbos die together. Although it is possible to see a body at the crash site, it is not overly graphic and there is no highlighting of there being any body parts. Most of the visualization of the dead revolves around the handling of corpses in blankets and the coffins. When seeing inside the morgue one can briefly see some bodies, but the angle is so wide that there is little detail. Further shots of the coffins when the lids are removed are done in such a way so that the contents cannot be seen. Those *izoku* who have seen *Shizumanu Taiyō* have commented how good they thought it was, how well recreated the scenes in Fujioka were, and how they spent much of the film crying.[60]

Chapter 7 questioned the non-acceptability of images of victims of plane crashes such as JL123 compared with the images often shown of victims of the bombings of Hiroshima and Nagasaki, for example. But why is it that we are accustomed to, and perhaps expect, graphic images of victims in dramatizations of these bombings, but would probably not expect or desire to see such images in a dramatization of a plane crash? Such images would probably deter some from going to watch the film, and so reduce its profitability and impact. But that point does not answer why the images in one are more acceptable than in the other, when, based on many of the original images I have seen, the content is not significantly distinguishable.

In addition to the dramatizations of the two novels relating to JL123 there have been other outputs. A low-budget independent film called *Osutaka-yama* was originally released in 2005. This film was directed by Fumiki Watanabe, who has often been the centre of controversy because of his anti-Imperial films, and uses the idea that JL123 was shot down as its premise. There have also been two plays, '*Akai tori nigeta...*' *(The red bird escaped)* and '*8·12 (Hatten ichini)' (8/12)*.[e] In 2005, TBS showed a docu-drama programme focusing upon the work of Hideo Fujita and how the CVR got into the public domain. In many respects it is a dramatization of Fujita's book – *Kakusareta Shōgen* (hidden evidence). Consequently, the main focus is upon discussing the cause of the crash and why Fujita believed that the official investigation got it wrong. The programme won the award for best docu-drama at the 11th Asian Television Awards and the 2006 award for best TV news programme by the National Association of Commercial Broadcasters in Japan.[61] Fuji TV's 2005 docu-drama, although also concerned with the cause of the crash, is much more focused on the human side of the story of JL123, and particularly concentrates on the experiences of the Miyajima family and how the 8/12 Renrakukai came to be formed.

What is important to remember about all of these productions is that they are as much, if not more, concerned with creating a product to be consumed as they are with 'the truth'. They may be repeated many times on television and many will watch the DVDs, meaning that their influence can continue for many years, as has been seen with the discussion of Kennedy's assassination, for example.[62] This influence can reinforce ideas on those who have already seen the output and introduce the ideas to those who have not seen them before. If the output is the latter group's first exposure to discussion about the event, there is a possibility for the output to be particularly influential. Heinze has suggested that in Japan, it was television rather than movies, as was the case in the US, that has been the means to tell the national story, and TV consumption is still higher than in many other countries.[63] In this way, the decision for NHK to make, broadcast and subsequently repeat *Kuraimāzu Hai* is significant. The showing on TV of the film version – which first happened in August 2009 on a Saturday just before the anniversary – and *Shizumanu Taiyō* may be of greater significance than how they performed in the cinema. If we look at the TV ratings according to Video Research Ltd for the Kantō region, we see that the TBS docu-drama got 14.1 per cent of the share of the viewing figures, making it the top programme in August 2005 for its category, whilst the Fuji TV programme got 17.6 per cent of the market and was fourth in the drama category. The NHK dramatization of *Kuraimāzu Hai*, did not make the top ten when it was shown in part as there was a major sporting event being broadcast at the same time as the second part,[64] but the film version of *Kuraimāzu Hai* was the third most watched film of August 2009 with 11.2 per cent of the viewing figures.[65]

Global narratives on JL123

Let us now consider English language sources relating to JL123. The crash happened as the world was paying more attention to Japan's economic miracle and looking to learn about Japanese management techniques. Many books about Japan appeared as a consequence, but few discuss JL123, not even those that mention other scandals or issues relating to JAL or the airline industry in Japan. van Wolfren's book is perhaps the exception which proves the rule. Most books that contain details of JL123 are not books about Japan, but books about aviation accidents in general. However, only Job has a whole chapter devoted to the JL123 crash,[66] whist others, if mentioning JL123 at all, generally only give brief mention to the crash. The need to summarize means that there is a tendency for the information to be fairly consistent, with no mention of there being any questions marks about the cause of the crash. Barlay comments, in a footnote, that there are questions being asked, but gives the impression that there is little need to take such questions too seriously.[67] Bartlett appears to be alone in pointing out that questions raised about the official report may be more than mere conspiracy theories.[68]

Job's book included the last image of JL123 taken by Keiichi Yamazaki. This image was also used by the Australian Civil Aviation Safety Authority's *Flight Safety Australia* magazine when they ran an article, written by MacArthur Job and Steve Swift, when it commemorated the twentieth anniversary of the crash (see Figure 8.4). Compared with the one used in Job's book, a red circle had been added around the place where the tail should be. However, if one compares the image with the one shown in Figure 4.5, it is clear to see that the addition of a circle is not the only difference. The original clearly shows that there was more tail as the plane passed over Yamazaki. Indeed, to the layperson, it may not even be clear that there was a significant amount of tail missing. A cropped version of the *Flight Safety Australia* image was on the Wikipedia page for some time, but it clearly over-exaggerates the truth about the damage done to the plane and so could help to perpetuate misunderstanding about what happened to JL123.

This has not been the only problem with the Wikipedia page. One of the issues with sites such as Wikipedia is who is writing and editing the pages? In essence anyone can. Lennon and Foley suggest that the 'accessibility and relative "openness" of the Internet allow for the possibility of all types of opinion to be aired, including some which may directly challenge prevailing orthodoxies'.[69] This is a relatively positive view of the situation and which does not fit with my own experiences. Attempts to revise the section relating to the cause of the crash on the English Wikipedia page about JL123 were removed with comments about how the page should not contain 'conspiracy theories'. A further attempt to reword the page to point out the difference in contents between the Japanese and English page was also removed. It was because of the concern about the possible spread

Figure 8.4 JL123 narratives outside Japan

Reproduced with permission, Flight Safety Australia, www.casa.gov.au/fsa.

of inaccurate information through sites such as Wikipedia that the World Wide Web Foundation was created by the creator of the internet, Sir Tim Berners-Lee.[70] Internet pages can change overtime, which is why it is important that bibliographic references include the date on which the page was accessed so that a service such as archive.org can be used to find an appropriate version. The changing nature of the internet can be an advantage as it can respond to new information – whereas books, unless a revised edition is produced, are static. It is perhaps ironic then that at least one company has sought to make profit out of Wikipedia by publishing books using the internet pages, without any additional checks being made on the reliability of the information contained in those pages. One such book that covers JL123,[71] for example, does not contain any reference to any issues with the official cause of the crash, or refer to the note from Wikipedia itself, which was placed on the page in 2009, requesting that the text be expanded with more text translated from the Japanese page.

Although there has been resistance to update the English Wikipedia page about JL123 to include one of the main alternative theories about what might have caused the accident, one part of the JL123 story has gained greater acceptance among the English-speaking audience. As mentioned in Chapter 4, in 1995, former American serviceman Michael Antonucci had an article published in which he claimed an American forces helicopter from Atsugi Naval Air Station reached Osutaka-no-One a little over two hours after the crash and that they were preparing to winch crew down to the site before being ordered to base.[72]

Whilst there is little to corroborate Antonucci's story, it is one that was also included in the BBC documentary about the crash and was central to the argument that the Japanese were unable to cope with the situation.[73] There are problems with Antonucci's article. For example, it exaggerates the degree to which his own plane's involvement was not acknowledged by the Japanese and has incorrect information about the timing and nature of the Japanese SAR activities.[74] Kawamura also questions the validity of Antonucci's story.[75] However, Kawamura appears to have been working from an inaccurate translation of the Antonucci article, as he claims that Antonucci suggested that the helicopter was to use another ridge due to the fire at ridge on which the plane had crashed.[76] There is no text such as this in the original article. Yoneda devotes a whole chapter to studying the Antonucci article, and although notes that parts of it are accurate, it questions some of the details.[77] Whilst the English-speaking world has generally accepted the Antonucci account, the Japanese appear to be more sceptical.

I believe that we have seen a Galápagos effect in relation to JL123. The Japanese-speaking world and English-speaking world have two largely unconnected versions of the JL123 story. Although the origins may have similarities – particularly the AAIC report – after that point they have evolved separately. This is not to say that there is a single view within either the Japanese-language world or English-language world. The article by Antonucci, for example, whilst being seemingly accepted without question by the English-language world was also accepted by parts of the Japanese-speaking world, particularly the 'conspiracy theorists'.

One consequence of having the two language groups living in relative isolation is that English-language documentaries seem to be largely based on the work presented in the English-language books that cover JL123. Although they may contain original interviews, the thrust of the programme suggests that little attention has been paid to the contents of the Japanese books. The consequence is a bit like a game of Chinese Whispers, whereby errors and alternative theories become an accepted part of the story and included in later documentaries. On top of this, the programme has a degree of poetic licence to make it good viewing material. The result may be that what is portrayed is somewhat different to what actually occurred. For example, 'Out of Control',[78] shown on the National Geographic channel in many countries, probably exaggerates the decompression and the panic on the plane, and also helps to perpetuate the myth that the captain controlled the direction of the plane through differential power to the engines on either side of the plane, which according to the official report,[79] did not happen. As discussed above, the power of the moving image, let alone the propensity to trust reputable companies such as National Geographic and BBC, means that these programmes may be treated as 'fact' and 'the truth', leading to the discounting of any other interpretation of what happened as a conspiracy theory. A further significance of this is that many of these programmes will be exported to other countries around the world, perhaps more than Japanese programmes will be, meaning that their influence reaches beyond the English-speaking world.

As well as the more formal outputs described above, YouTube has become a popular source for media relating to almost any subject. In relation to JL123 there are over 250 videos, posted by around 100 users, with the number continuing to increase. The type and quality is variable. Some are copies of some of the documentaries, including Japanese ones, mentioned above. Some have uploaded whole sections of the TV coverage from the evening of 12 August. Whilst shorter clips of this type may have been taken from other programmes, others are of such length and quality to suggest that the user videoed the coverage in 1985, has then digitized it, and then decided to upload it to YouTube. Although most users did not respond to messages about why they posted such videos, those that did appeared to be concerned with ensuring JL123 is not forgotten or so that the truth may be better know. In relation to the last point, there are many videos, with many from a limited number of users, which present material related to the idea that JL123 was shot down. A number of videos also provide some form of animation together with sounds taken from the CVR. Of the videos posted by those outside of Japan, other than clips from English-language documentaries, the other main type of video is where the user has attempted to recreate JL123's flight using software such as Microsoft Flight Simulator. As with the Wikipedia page, one has to be concerned about the influence of some of these videos given the mistakes contained with some of them. But who is viewing these videos? Lennon and Foley argue 'that it is unreasonable to separate the viewing public which assimilates these media commodities and those who seek tourism experiences'.[80] In that respect, perhaps those visiting

YouTube sites about JL123 should be treated the same as 'dark tourists' going to Osutaka-no-one (see below).

Continued interest in JL123

The number of websites and YouTube videos points to there being significant continued interest in JL123. But what is the basis of this interest? Trying to find the answer to this question was one the starting points for my research. To help with discovering the answer, I set up an online survey. This survey ran for approximately three years and there were both English and Japanese versions. Although the survey pages were accessed over 3,500 and 19,000 times respectively,[81] there were only a total of 169 valid submissions of the English survey and 143 of the Japanese version (see Chapter 1 about problems with submissions). Despite the somewhat disappointing numbers, they were sufficient to suggest answers to the question about which issues relating to JL123 make the greatest impression upon the respondent. Some of the suggested answers, as expected, had a relatively low score due to the polarizing nature of the issue. For example, knowing somebody on JL123 or the possibility of having been on JL123 themselves was a highly significant reason for interest in JL123 where this condition applied, but of no significance for those where it did not. Many of these questions were included to help check the background of the respondent and that the survey was being filled in correctly and so there is no need to summarize the results here. Figure 8.5 summarizes the most significant answers.

As can be seen, the results for many of the significant reasons for interest in JL123 do not vary significantly between the two groups. However, we do see that the respondents to the Japanese survey found it more significant that there is so much interest in the crash more than 20 years on, that the crash happened in Japan, that it was operated by the national flag carrier, JAL, and that Kyū Sakamoto was amongst the victims. However, it is also noticeable that many more respondents of the Japanese survey commented that their interest was due to concerns about the official cause of the crash. This would certainly appear to confirm what has already been discussed in this chapter in relation to the nature of the publications about the crash. A similar survey about the *Titanic* would probably find comparable results, with the uncertainty about exactly caused the sinking being of interest to many.

The survey also found that, of those who answered the Japanese version and who were alive at the time of the crash, most had a clear memory of where they were when they first heard news of the crash. Although I expect most who answered the survey have an interest in the crash so this finding is not surprising, I have found that many Japanese I have spoken to about my research, but who do not have a particular interest in the crash, can also remember where they were at that time. In this respect, JL123 would appear to be a Japanese equivalent of the deaths of JFK and Lady Diana, or the 9/11 attacks.

Analysis of what searches took people to my websites[f] also reveals some of the main interests that people appear to have. Popular searches include 'JAL123 passenger list', 'JAL 747 crash messages', 'Japan 1985 Mt Fuji crash', 'Kyū Sakamoto note to wife', and 'JAL 123 survivors where are they now'. These searches indicate not only what the interests of people are, but also a level of ignorance about the crash. For example, JL123 neither crashed on Mt Fuji nor was any *isho* by Kyū Sakamoto found. Looking at the dates of when my websites were accessed, the day with the highest number of hits is 12 August. Similarly, a search of 'tweets' including the term 'JAL 123' made on the social network Twitter reveals that the vast majority are in August, and particularly on 12 August, when it was one of the most popular topics. But what makes people think of the crash on this date? In Japan, it may be the reporting of the anniversary in the news. However, many of the tweets and accesses to my websites come from other countries where the reporting of the anniversary is rare.

Although there were a number of *izoku* who completed my survey, the opinions of non-*izoku* in relation to one question; whether they would like to visit the crash site, Irei-no-Sono or the Safety Promotion Center (see below) were also

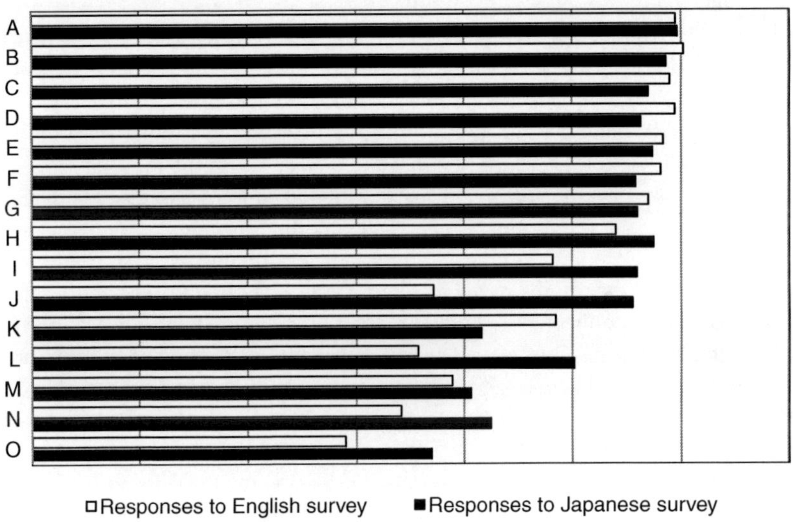

Figure 8.5 Reasons for interest in JL123

Figure by the author. A: That so many people died; B: That a US military helicopter's help was apparently rejected; C: That JL123 stayed in the air for over 30 minutes after the explosion; D: That there could have been more survivors had the SAR been quicker; E: That the SAR took more than 12 hours to reach the crash site; F: That it is the world's largest single plane crash in terms of fatalities; G: That there were survivors; H: That some wrote *isho*; I: That there is so much interest in JL123 over 20 years on; J: That the official reason for the crash may not be correct; K: That the accident was bound to happen one day if the official reason for the crash is correct; L: That it was in Japan; M: That it was a Boeing 747 jumbo jet; N: That it was operated by JAL, the national flag carrier; O: That Kyū Sakamoto was one of the victims.

enlightening. This interest in visiting places connected with death is commonly referred to as 'dark tourism', which appears to be a growing phenomenon, although pilgrimage as one of the earliest forms of tourism was often also concerned with death.[82] Lennon and Foley suggest that central to the growth of 'dark tourism' has been media as many of the destinations relate to events which have occurred since the invention of cameras, radio, television and modern communications.[83] This tends to mean that the sinking of the *Titanic* is seen 'rather arbitrarily' as being the 'starting point' for 'dark tourism'.[84] Lennon and Foley say that there are three critical features within 'dark tourism' – global communications creating the interest; the incident in question introduces 'anxiety and doubt' about modernization; and, there are opportunities for 'commodification' and 'a commercial ethic which (whether explicit or implicit) accepts that visitation (whether purposive or incidental) is an opportunity to develop a tourism product'.[85] The first two elements are clearly evident in the case of JL123, whilst the books and DVDs related to JL123, as well as the *omamori* which can be bought at Irei-no-Sono, can be seen as 'commodification'. Lennon and Foley also wonder whether 'dark tourism' can only happen in large urban areas, but acknowledge that if the scale of the tragedy is large enough, it might be possible at other locations, such as Auschwitz.[86] I would suggest that in addition to access from an urban areas, what might be needed is something iconic at the place where the tragedy happened. Kennedy's assassination has the book depository and grassy knoll, Auschwitz has the railway lines leading up to the building, Hiroshima has the A-bomb dome, and the Fukuchiyama Line derailment has the block of flats the train hit.[g] Perhaps *Titanic* is exceptional since the site of its sinking has nothing there and is difficult to access. For JL123, the mountainous crash site, the U-shaped cutting, and even Irei-no-Sono can be iconic. Similarly, physical things related to the tragedy can be important. In this respect, the items recovered from the wreck fill the void left by not having a place connected to *Titanic* to visit. For JL123, there are still parts of the plane at Osutaka-no-One as well as larger parts at the Safety Promotion Center (see below).

Although I use the terms 'dark tourist' and 'dark tourism', I think that in many cases there is a fundamental flaw with them. Walter points out that little is known about the motives of such tourists and also questions the need to do so,[87] unlike many others who contributed to the same book. However, I suspect that a large percentage of 'dark tourists' go to 'dark sites' in spite of, rather than because of, the association with death at that site. Instead, I believe it is the symbolism of the place or event itself which is significant. For example, Formula One fans will travel at non-race weekends to see the tunnel at Monaco, Beatles fans will go to the Cavern club, and film fans will go to see where particular scenes were filmed. Visiting these places can provide a physical connection with something significant to them. Most of our lives are relatively mundane, but there are moments of particular happiness and sadness. Some of these are connected to places due to the imagery or text in the news, in a film or in a book. It is not surprising, therefore, that some will seek out these places as it may help to recall happy memories, come to terms with sadder memories or merely to connect with something out of the ordinary. Indeed, Reader[88] points out that this last one has been a motivation for many pilgrims and Chapter 6 has already noted that the distinction between pilgrim and tourist is often very blurred. Visitors to Japan may seek out the sites which are familiar to them from images in films, such as the scramble crossing in Shibuya, as much as those recommended in guide books. Whilst 'dark tourism' may exist, and 'dark sites' may need to be looked after differently to other tourism-related sites, I think it is naïve to suggest that the fact that people died there is the overriding reason for a site's interest. We must not overlook the significance of nostalgia, childhood memories and the desire to connect with the significant.

Whilst the numbers of 'dark tourists' to Irei-no-Sono and Osutaka-no-One are increasing, it is hard to be certain of their motivations for going. Websites written by visitors tend to describe the visit rather than motivations for the visit. As such visitors are rare around the time of the anniversary, I have had limited opportunities to talk to any, but perhaps one should keep in mind the motivations of pilgrims more generally, as discussed in Chapter 6. Further, as Creighton points out 'Travel that is edifying as educational, such as a journey into Japanese history, provides a similar cloaking of legitimacy to recreational travel, the legitimacy of which few would ridicule or question.'[89] The educational element of 'dark tourism' may be particularly significant in Japan, in other words. Naturally, anybody can visit Osutaka-no-One and Irei-no-Sono, but generally non-*izoku* visits are rare on the anniversary, with signs discouraging non-*izoku* from visiting Osutaka-no-One until after 14:00 on the 12 August.[h] Visitors can pick up a map of the crash site after starting the climb, but information is limited. If you are trying to find a particular *bohyō*, you will not able to do this unless a JAL employee is there as they have a list which has these details. I believe that the number of 'dark tourist' visits is likely to increase in the coming years as *Kuraimāzu Hai* and *Shizumanu Taiyō* reignite interest in JL123. Those who maintain Osutaka-no-One have commented to me on the number of such visitors has already been increasing in number. Such visitors are generally recognizable as they often only visit the area around *Shōkon-no-Hi* and if they have brought flowers, will leave them there. Should they visit a specific *bohyō*, it tends to be that of Kyū Sakamoto.[i] Indeed, Lennon and Foley note that many 'dark tourism' sites and its commodification relate to events in which famous people are being remembered.[90] Similarly, for example, when a singer dies it is not uncommon for fans to buy their songs and albums. In 2007 I was one of those who bought a greatest hits collection of songs by Zard after the tragic death of the singer Izumi Sakai. This is not a phenomenon restricted to the world of pop music. Pavarotti's CDs also enjoyed a similar revival in sales after his death in 2007. Even House Foods saw improved sales of its goods, such as instant noodles, following the death of its president on JL123.[91]

However, there is a significant difference between JL123 and many other locations of 'dark tourism'. Many of the other sites around the world have been maintained as they looked at the time of the event. Others, such as the A-bomb dome in Hiroshima or Auschwitz, have the problem that 'the place of death is itself dying',[92] so consideration is needed as to how to better maintain it. But in the case of JL123, the trees have grown back as was discussed in Chapter 6.

Visiting the site can help one to begin to understand the scale of the crash which photographs and words struggle to do. Seeing the U-shaped cutting on the next ridge, walking around the extensive site, walking up and down the steep slopes allow the visitor to recreate the events in their head, despite the site looking very different now. That there are question marks about what happened provide further ammunition for what to think about whilst there. It is this range of issues which may be part of the appeal, just as West Sellars and Walter claim it is for those who visit Little Bighorn or the book depository from where Kennedy was shot.[93] I also think that we should not overlook the importance of stories – whether from family members, books or elsewhere – upon children and how, as adults, they may become 'dark tourists' as a means to rekindle a memory and experiences from when they were younger. Similarly, there may be a desire to visit the place, the images of which are familiar to them from news, documentaries and movies, as it is 'familiar', but at the same it may also be that they are wishing there was *no reason* to go there and that a visit may help 'wish it away'.[94]

Lennon and Foley point out that places of 'dark tourism' are likely to involve 'some element of storytelling',[95] although it has also been noted that some of the former Nazi concentration camps have limited signs and information so as not to restrict 'reflection and contemplation' and 'understanding and appreciation' through the over-use of 'historical/literal interpretation'.[96] Although there is a small 'museum' at Irei-no-Sono, what is there is limited in nature and barely tells a story. Much more powerful is the Safety Promotion Center established by JAL in 2006. The Center is similar to the accident museum found in the JR East training centre,[97] although JAL did not visit this one.[98] Although the Center is primarily for JAL employees to visit as part of their training, there are visitors from other companies and the general public (see Appendix 2 for details), and can be considered a possible 'dark tourism' destination. As Lennon and Foley point out, the place where 'dark' events are interpreted may be at another site than where the event took place.[99]

Although the Safety Promotion Center is not only about JL123, JL123 is certainly its most significant part. As well as TVs and charts showing information about the crash, there are part of the plane – such as the broken rear bulkhead (see Figure 8.6), part of the tail fin, and some of the seats from inside the plane – as well as some of the victim's *ihin* (see Chapter 9). Of course, with some raising questions about the cause of the crash, the way in which the Center deals with this topic is significant. Lennon and Foley point out that it takes time for a place which interprets the event in question to be formed, coming after the erection of memorials, for example, and that such places are a 'critical' part of 'dark tourism'.[100] The Safety Promotion Center's establishment fits with this, although I wonder whether there is still more to be done. That the Center is housed by JAL is likely to mean that there are limits to the 'interpretation' of the crash, and this may mean that Osutaka-no-One remains a better place for the 'dark tourist' to consider their own ideas about what happened.

That the Center is rather small, relatively hard to access and in a place essentially unconnected with the crash are other issues. But there are limits to what can

Figure 8.6 The Safety Promotion Center

Picture courtesy of JAL International.

be done. It is unimaginable to have a display which recreates the flight. Although such a simulation, if done properly, may have its uses there is a danger that the real message of the Center (or any new museum) could be lost as the display may become too much like a Disney ride or merely not be able to satisfactorily recreate the awful conditions as the plane was in a phugoid motion and Dutch roll, just as London's Science Museum's recreation of the Great Hanshin Earthquake misses the mark.[101] Similarly, it is unimaginable that anyone would put on real flights that recreate the flight path of JL123 as has been done with cruises that recreate parts of the *Titanic*'s journey and end with a memorial service above the location where the ship sank.[102]

Just as the *hibakusha* of Hiroshima have been central in determining how the bomb should be remembered, and that this has changed over time until by 'the early 1990s the *Shiryōkan* finally had become a full-fledged history museum (*hakubutuskan* or *chinretsukan*)',[103] so the *izoku* are key to the contents of the Safety Promotion Center. Indeed, the *izoku* have even been on fact-finding trips to Hiroshima to see how items are displayed and the bomb is commemorated to aid them with thinking about how they would like JL123 to be remembered in the Safety Promotion Center.[104] In particular the *izoku* placed pressure on JAL to display more *ihin* and some *izoku* have become involved in speaking directly to JAL trainees visiting the Center so that they can better understand their experiences.

Nakamaki suggests that the 'memorial monuments play a role in acknowledging death socially and producing collective memory. In other words, erection of a memorial monument is a method of social cognition, and it functions as a device for the transmission of memories.'[105] Any of the three locations related to JL123 could and should perform this role, but I wonder whether any of them are doing this as effectively as they could. I believe that it is important that more is done to enhance the experience of 'dark tourists' who visit Osutaka-no-One, Irei-no-Sono or the Safety Promotion Centre as will be discussed in the next chapter.

Part V

The legacy of JL123

9 Conclusions

JL123 present and future

18:56 on 12 August 1985 marked a new beginning for some, whilst for others their world stopped at this time (see Figure 9.1b). This book has covered a wide range of issues relating to JL123 and this chapter will highlight some of the key issues while considering the longer term legacy of JL123.

This study has emphasized the need to understand JL123 in context. An example of the problems can still be seen when, even in 2010, *The Japan Times* wrote of the memorial garden, Irei-no-Sono, being at 'the foot of the mountain' where the crash happened,[1] yet, Osutaka-no-One is over 20 km away from Irei-no-Sono. Although today some networks provide a limited coverage on Osutaka-no-One and within Ueno-mura, mobile communications are limited. But this is still better than it was in 1985. 1985 was still a time when mobile communications were in their infancy. While most households had colour TVs, there were far fewer channels and many of these ceased their broadcasts during the night. There were no dedicated news channels in Japan. It was in this world, that, during the summer when Japan was commemorating the fourtieth anniversaries of the bombings of Hiroshima and Nagasaki and was preparing for the anniversary commemorating the end of the war itself, that the JL123 crash occurred. JL123 itself was a specially adapted Boeing 747 for use on Japan's heavily used domestic routes. Being a flight at the end of the day, the passengers were a mixture of people, and many were returning to their ancestral homes for Obon, the festival of the dead. Others were taking advantage of the preparedness of companies for holidays to be taken around this time to travel with their family. An understanding and appreciation of these points, discussed in detail in Chapters 2 and 3, are critical for comprehending the significance of JL123.

But this study is more than merely an historical account and analysis of events in 1985, including the response by the various actors involved and the impact it had upon those who lost loved ones (as discussed in Chapters 4 and 5). The JL123 story continues. Acts of remembrance and memorialization take place around the time of the anniversary and are covered extensively in the media, as discussed in Chapters 6 and 7. These acts tell us much about Japanese attitudes towards death and religion. But why does the media cover this story? Why are there so many books, documentaries and even movies about the crash? The answers to these questions, presented in Chapters 7 and 8, point to JL123 being special in some

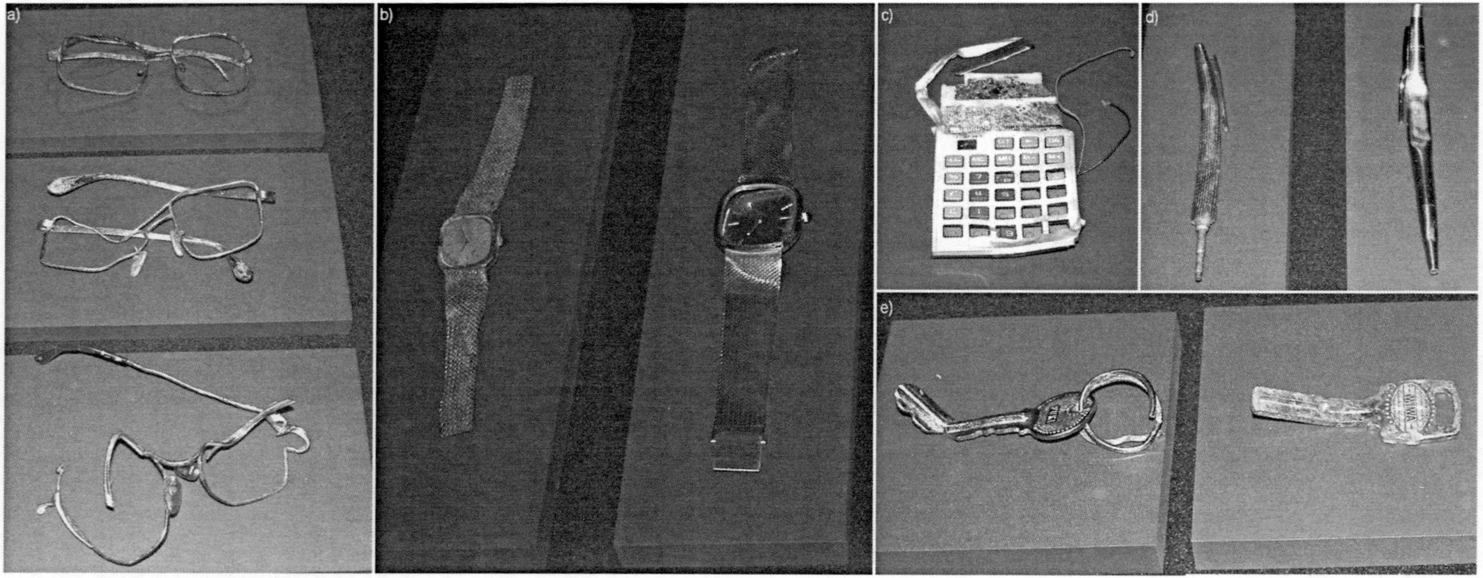

Figure 9.1 Ihin at the Safety Promotion Center

Pictures courtesy of JAL International.

ways. This special position, however, comes with an additional responsibility. Remembrance of JL123 is not just about JL123 itself. It is about the importance of aviation safety and traffic safety in general. It is about the importance of life. In that respect the story of JL123 has relevance far beyond those directly connected with the crash. Whilst the number of those interested in the crash and visiting locations related to the crash site may be increasing, perhaps more needs to be done to ensure that the lessons of JL123 are learnt by more. To do this means ensuring that there is agreement about the facts relating to JL123. And here there is a problem, for there are still questions about the facts relating to the cause of the crash – at least among many Japanese, even if not so much globally.

The cause and response

There are two areas where lessons could be learnt. First, there are areas related to the design, maintenance and operation of the plane itself. To learn the lessons here one has to be clear about what the cause of the crash. The 8/12 Renrakukai, set up by *izoku* of the JL123 crash (see Chapter 6) wants to ensure that a crash like this never happens again,[2] but how can a second time be prevented if we cannot be sure that we know what happened the first time? The second area where JL123 may have useful lessons relates to response to such an accident, in terms of areas such as the SAR activities and the support given to *izoku*. 'Lessons' does not necessarily mean that what happened in 1985 was wrong, as there may be positive lessons which need to be remembered, but the areas where there could be improvements appear more numerous.

As stated in Chapter 1, when I began this research I had no intention of discussing the cause of the crash in detail. My reasoning was that, having read English sources and seen English documentaries, there was no reason to. None of these sources had made any suggestion that there may be any problem with the official investigation's conclusions. Once I began reading Japanese sources, however, I realized that there were questions about the cause and that, in writing the first English book solely about JL123, it would be inappropriate if I did not discuss some of the points raised in these Japanese books. However, to suggest that I could reach a correct conclusion about what happened would be folly. Although I have learnt much about aircraft safety, engineering and crash investigation in the course of this research, I am not an expert. However, I am deeply concerned that the original investigation was not done as thoroughly as it should have been, that the report was poorly written and that there is a possibility that the conclusions of the original investigation survey may have been wrong.

Chapter 4 highlighted the concerns raised in many Japanese texts about the length of time that JL123 spent above 20,000 ft – an altitude at which those without breathing equipment would have suffered hypoxia. The cockpit crew did not use their oxygen masks. Although the report says that the limited conversation between the crew was an indication of them suffering from hypoxia, other evidence, seemingly not considered by the AAIC, but discussed in this book, suggests that there were other reasons for the limited conversations. In essence the crew were so busy fighting to control the plane, using great physical effort due to the hydraulics of the plane having been completely lost, that conversation, was both impractical, and given that the need to focus on the task at hand was a priority, was perhaps unnecessary.

The key point is that if the whole plane was not suffering from complete depressurization, it may mean that there had been no been rapid depressurization. If this is the case, then the chain of events, as described in the AAIC's report, could not have taken place. Perhaps, the damage to the rear bulkhead happened in such a location that the 'rapid' part of the depressurization only occurred within the section above the ceiling of the cabin section, as those such as Katō have suggested.[3] If this is the case then perhaps it accounts for why those who wrote *isho*, the survivors and the cockpit crew do not make reference to any of the typical features of *rapid* depressurization. As discussed in Chapters 4 and 5, Yumi Ochiai, one of the survivors who has spoken most about her experiences, and two of the *isho*, do make reference to features that are common to depressurization, but not rapid depressurization. But if the rapid depressurization did occur in this way, the AAIC report does not clarify that this is what they concluded. It does not clarify whether such a scenario is consistent with what the *isho*, the survivors, and the cockpit crew did and, perhaps more significantly, did not report.

But does science support the AAIC conclusions? Certainly Katō presents a convincing argument to show how the depressurization may have occurred in accordance with the AAIC report.[4] Frank Taylor, who has investigated many aircraft accidents, also thinks there would be a possibility that the cabin pressure would go down quite slowly, as the hole in the bulkhead may be relatively small whilst the cabin volume is particularly large on a Boeing 747.[5] But if the hole was small enough that depressurization of the cabin was relatively slow, could it have been large enough to allow enough pressurization to enter the unpressurized tail section leading to its structural failure? The AAIC is clear about how much pressurized air would have to have entered into this area and in what time period as calculations can be made based on the strength of the structure, the volume of the air in the tail section, and using information taken from the DVR. However, all of this works on one key assumption: the pressure relief door worked properly. The pressure relief door is meant to be triggered once the variation in pressure reaches a critical point. But it is limited in size. Only a certain amount of air can pass through it in a particular time. The AAIC worked on the assumption that the pressure relief door function properly, and so was then able to calculate how much additional pressurized air must have rushed up the tail section as a consequence.

However, the AAIC report itself confirms that its own tests found that it was 'impossible to clarify' whether the door, 'which was discovered in the vicinity of the crash site', 'opened in the early stage of the crash'.[6] Yet, the AAIC says that 'the possibility is considered high that the door opened because of the design to open... and the condition of damage to the door'.[7] Are these really meant to be considered valid arguments? That it was designed to open does not mean that it *did* open. Although more detail is not given in relation to the 'damage to the door', given their own tests were not able to confirm whether it operated properly or not, it seems a large, significant jump of faith to suggest that it probably did open. However, it is perhaps not a surprising jump given, as discussed in Chapter 4, the AAIC seemed happy to conduct an investigation and reach a conclusion which was consistent with the information provided to it by Boeing.

But I am not convinced that the pressure relief door did work properly. What could have been the chain of events in such a situation? If we assume that all the information given about the faulty repair to the rear bulkhead after its 1978 accident, as discussed in Chapter 4, is correct, then the rear bulkhead still could have failed. However, rather than requiring a significant rupture, there would only needed to have been a small gap. Gradually, over the course of 12 minutes, pressurized air would seep into the tail section. Eventually it would reach a point where the structural integrity of the tail would fail. It may be that as a consequence of this that additional damage would occur in that section of the plane, including to the rear bulkhead. However, the gap between the cabin section and the tail section may have remained relatively small. This could account for why there were no accounts of rapid depressurization. It could also account for why the cockpit crew were able to function without the use of oxygen masks. The ultimate conclusion about the cause of the crash – that it was due to fatigue in the rear bulkhead due to its faulty repair – remains the same. All that has changed is the assumption about whether the pressure relief door functioned properly or not. But doing this is not without its consequences as will be discussed below.

During the original investigation a hearing took place to discuss the findings of the AAIC to that point. This hearing is best remembered due to the comments by one of the speakers provided by the AAIC. Prof. Keihachirō Shimizu argued that the cause of the JL123 crash was due to three factors: first, that Japan is a mountainous country; second, societal restrictions on Itami Airport; third, that Japan is reliant on America's aviation industry.[8] Some *izoku* were so incensed by what they heard that they stormed out of the hearing in protest. At one level I can understand what Prof. Shimizu was trying to suggest. After all, the cause of the sinking of the *Titanic* was not an iceberg as icebergs by their very nature are supposed to be in the sea whereas humans, as land-based animals, are not. The *Titanic* was there due to the pursuit of luxury and the emigration to the USA. The lack of lifeboats, the design of the ship and the actions of the captain are secondary issues. Similarly, JL123 is not so much about the plane itself as the need to have such planes. The phenomenon of *tanshinfunin*, travelling for Obon, the pursuit of leisure in Tōkyō, the demand for mass transport which leads to the use of such high capacity jumbo jets being used on a domestic route and the pricing of tickets which do not reflect environmental costs so that the shinkansen does not always appear as advantageous as it could, are all factors. But these factors are generic. They do not explain why some planes do not crash and why one plane did. This is the point which Prof. Shimizu's argument appeared to ignore. In relation to JL123, there were many more specific issues as to why the plane had its accident. The first of these was being set in motion as early as the mid-eighteenth century,[9] when a tree began to grow on the slopes of a ridge of a mountain near the borders of what would become Gunma and Nagano prefectures. By 1985 this tree was around 14 m tall[10] and was hit by the right wing of JL123 as it pulled out of a steep dive and struggled, but failed, to clear a ridge, after which it would continue powerlessly towards another ridge where it crashed. However, it seems that the most crucial causal factor specific to JL123 was the accident that occurred in 1978 and that this apparently led to a mis-repair of the rear bulkhead. Standing next to the bulkhead in the Safety Promotion Center it is incredible to think that this piece of equipment was probably key to the death of so many people.

But there are many alternative theories as to what led JL123 to crashing, as was discussed in Chapter 8. Although I am not convinced by the idea that JL123 was shot down, some of the issues raised in this study do leave me wondering if there is still more to some of these than most would credit. The photograph taken of an object outside JL123 by one of the passengers, the questions raised about who was at the crash site during the night and why the SAR appeared to be so slow are all fuel for an alternative 'conspiracy' theory. But surely if any of this was based on fact, there would be some evidence, some clue to its validity. There are a number of websites which suggest that the photographs taken from inside JL123 do this as they show that the plane was lower than the altitude indicated on the FDR.[11] They also argue that the information provided by those who witnessed the explosion as it flew close to the Izu Peninsula points to the plane being lower than suggested. Could it really be that the FDR which has been made public is a work of fiction? It is hard to believe. But Aoyama's comments about the rear bulkhead make it hard to believe that the bulkhead on display at the Safety Promotion Center is the one from JL123. She points out that the press were kept away from this key piece of evidence from the afternoon of 17 August as reported in the *Jōmō Shimbun* on 18 August. This was after the press had already photographed it whole and how it was reported that it had been cut up into five pieces on 15 August.[12] Is the damage to the bulkhead on display at the Safety Promotion Center consistent with that seen when it was photographed on Osutaka-no-One? That it is in two pieces rather than five and signs of it having been cut up and rejoined are not immediately apparent certainly does little to negate Aoyama's concerns. One of those who had an opportunity to see key evidence for himself has described what happened to JL123 as 'murder'.[13] What did he mean? Is this merely a comment about the tardy SAR and how more lives could have been saved, or is he aware of some evidence that confirms the suggestion that JL123 was shot down? For now he will not tell his full story, but it may come out in due course as he has documented it and placed it in a safe to be opened following his own death.

Whatever the actual cause, there are clearly questions which remain unanswered. But can these questions be answered? There may still be key evidence relating to the cause of the crash in Sagami Bay. Although a survey was made in 1985, as discussed in Chapter 4, this survey may not have been sufficient. Hooper McCarty and Foecke suggest that 'In seawater, the skin of aircraft will last only a couple of months before being eaten away if it is not protected with a highly specialized coating.'[14] If this is the case then the key evidence may have even been gone by the time that the original survey was done. However, if the paint and other coatings have protected the parts, then vital clues to the cause of the crash might still be waiting to be found. As the Japanese government has so far not been inclined to reinvestigate the cause of the JL123 crash due to the lack of new evidence to warrant such an investigation, it would appear that it may require someone else to try to find the evidence. Just as James Cameron's visits to the wreck of the *Titanic* were key to learning more about the ship,[15] those who want a reinvestigation into the cause of the JL123 crash might need a wealthy individual to make the first move.

Ultimately I believe it is necessary for there to be a reinvestigation of the crash. I am alarmed at the number of question marks raised about the official accident report based on logic, evidence which had been apparently overlooked or ignored, and on engineering grounds. The result is that while the official report may not have got it wrong, I am extremely concerned that they did not get it right, and am concerned that they did not get right in the right way. There are many questions that still need answers. Some of these questions are unanswerable as key people, for example the cockpit crew, are not alive to tell us their thoughts. Other evidence could still be out there. Like the *Titanic*, it could be many more years before the probable facts are known. But I am not convinced that the official report's account of the accident is probable. In relation to Sherlock Holmes' famous adage that 'once you eliminate the impossible, whatever remains, no matter how improbable, must be the truth', the official report did not even eliminate the probable, let alone the impossible. What remained is itself improbable, and is questionable as to whether it was possible.

A further concern about the original investigation was the influence of outside 'interested' parties – such as Boeing. How could this happen? Perhaps we should not be surprised. After all for the certification of the Boeing 777, '95 per cent of the work in certifying the plane was outsourced to Boeing itself'.[16] This perverse system is equivalent to me asking my students to set and mark their own exams. I cannot imagine anyone, other than the students themselves, thinking that this would be acceptable. And yet, this system is allowed to prevail in a world where lives rather than marks are at stake. In relation to JL123, it seems remarkable that Boeing could have had so much influence then, but given that even in the more recent case of the Fukuchiyama Line derailment there was inappropriate influences being attempted by JR West,[17] there is clearly a problem with the way investigations have been conducted in Japan.

Although changes have been made, and even some who feel that a reinvestigation into JL123 is necessary, believe that these will help,[18] perhaps the changes have not gone far enough. While the AAIC was comprised of non-experts in crash investigation, housed within the Ministry of Transport, the new body is the Japan Transport Safety Board (JTSB), established in 2008 through the amalgamation of the Aircraft and Railway Accidents Investigation Committee (ARAIC)[a] and Japan Marine Accident Inquiry Agency (JMAIA). But I believe that this does not go far enough as the JTSB is still housed within the Ministry of Transport, has a relatively small budget and small number of staff. Although the number of aviation accidents in Japan is small, Japan could, and I believe should, have the provisions to have a specialist team for investigating accidents.

This team could become a resource for other nations which do not have the capability to have their own specialist teams. They could also be called upon for conducting investigations into any accident, anywhere in the world, that a Japanese national has been involved in. The British Air Accidents Investigation Branch (AAIB) works on this basis, drawing upon Annex 13 of the International Civil Aviation Organization (ICAO).[19] The AAIB is highly respected in Japan,[20] and, given the comparable scale of the countries' aviation interests, I would suggest that Japan would do better to look to learn from the UK's AAIB than the USA's NTSB.

I would also propose an additional alteration. In terms of formulating the final report, the panel should be made up of seven people. One of these should be somebody who is independent of the JTSB, but who has expertise in aviation matters. They should be able to demonstrate independence from the government, the manufacturer of the aircraft involved in the accident, and the carrier involved. This person would be appointed as the chair of the panel and would have to be approved by a majority of the panel members. In terms of the rest of the membership of the panel, I would propose that three would be representatives of the team that conducted the main investigation – experts in accident investigation – and that the other three be representatives of the *izoku*. I would hope that in this way the final report would be written in a language which would be understandable for those who need to understand it, usable for those who need to use it, and based upon detailed, well investigated and impartial research.

When an accident occurs there are lessons to be learnt. This is one of the primary reasons for having an investigation and a report. Following the JL123 crash, and indeed prior to the publication of the final report, some changes were made to the design of the Boeing 747. For example, a new hatch was put over the hole leading up the tail structure. Had this been in place in the case of JL123, then although there would have been some structural damage, most of the stabilizer would have remained. Second, valves were installed so that it should no longer be possible for all hydraulics to be lost. The FAA noted in a letter to Boeing which was obtained by one of the lawyers representing some *izoku* that 'it is now believed that had there been protection for one or more hydraulic systems, the airplane could have been landed with little or no crash damage'.[21] Why was it that these two design problems were not discovered during the 747's initial design? Similarly, why is it that the 747 and other similar planes are still able to fly without satisfying Section 2.1.3 of the British Civil Aviation Regulations, as discussed in Chapter 4? Whilst it would seem that some of the lessons of JL123, assuming the official report's 'probable cause' was correct, have been learnt, there are many more lessons which could have been learnt. As no autopsies were conducted, valuable information about the effectiveness of seat belts and seat design, for example, was lost. This point has been emphatically made by JL123 *izoku* and head of the *Kōkū Anzen Kokusai Rarī*, Dr Kawakita, who received an award from the National Air Disaster Alliance (NADA), a major US-based air safety group, in 2009.[22] It could take other accidents and more deaths to bring about further changes, through the process known as the 'tombstone imperative', as mentioned in Chapter 2.

Masato Yamaguchi 山口勝人 *(36)* · *Shizuko Yamaguchi* 山口静子 *(32)*

Perhaps what is still lacking is a proper 'safety culture'. I would not want to suggest that this is a problem that is either particularly prevalent or even more prevalent in Japan than other countries. But there clearly have been cases where safety has not been given due attention. Inside the aviation industry, there have been lapses that have led to accidents or nearly led to accidents. In relation to JAL, perhaps the two most infamous since JL123 are the cases of JO58, operated by JALways (a wholly owned subsidiary of JAL), and the near-miss between JL907 and JL958. In the former case, which happened on 12 August 2005, on the twentieth anniversary of JL123, some parts of an engine came off after the plane took off from Fukuoka Airport. There were no fatalities and the probable cause was found to have been erosion of the turbine blade in the engine.[23] In the second case, the problem was due to poor instructions from ACC, including the wrong flight number being used during one of the communications. In the end the two planes – with a total of 677 on board – came about 135 metres from hitting each other. As it was the evasive action led to 99 injuries.[24] There have been problems at other airlines too. For example, in 2010 the Ministry of Transport reprimanded Skymark Airlines for overruling one of its own captain's safety concerns,[25] and in 2009 a misheard flight number nearly led to a collision at Itami Airport.[26] Whilst we can consol ourselves that such cases are rare, most, if not all, should be avoidable, and so the question is to why they happen at all. Can a system be designed that compensates for human errors?

But the state, manufacturers and airline companies are not alone in sometimes appearing to not give due concern to safety. Many air plane passengers will at some stage drive cars. How many of these have never gone faster than the speed limit, never forgotten to indicate, never forgotten to use their lights, never driven whilst over the legal alcohol limit, never jumped a traffic light, never driven whilst holding and talking on a mobile phone, or never driven a car which was not totally road worthy? I suspect the figure is zero. I suspect the number who commit at least one of these offences every day is also depressingly high. In relation to air travel, in 2010 the skies of Europe were empty for several days due to volcanic ash billowing from an active volcano in Iceland. As the days passed and people became more desperate to go about their daily lives and as the economic impact began to affect more and more in not just Europe but around the globe, so the demands for flights increased. It was pressure from passengers, and also some airline companies, which ran their own test flights, that led to the resumption of flights. But were these test flights safe for the crew on the plane and for those living in their flight path? Everything has its price, even safety. With the drive to find cheaper flight tickets, I wonder how many people would be prepared to pay an additional amount in order to cover the expense of additional safety provisions on a plane.

What if there were a reinvestigation into the JL123 crash and a different conclusion was found? That should be the end of it. However, we have to remember that there were two investigations back in 1985. There was the AAIC investigation relating to the cause of the crash and there was also Gunma Prefectural Police's investigation. In many respects, as discussed in Chapter 4, the latter took primacy over the AAIC investigation. The police investigation was to ascertain whether a crime had been committed or not. Would a new, different conclusion mean that Gunma police would look to reconsider their position? In theory, there answer should be 'no' due to Japan having a statute of limitations. However, there have been discussions recently about the period of the statue of limitations being extended. Such an extension, or even abolition, could mean that Gunma police would also have to reconsider its position. Until I served on a jury dealing with allegations dating back over 30 years I found it hard to appreciate the need for a statute of limitations. Considering JL123, I suspect that if an investigation were to suggest that somebody should be prosecuted, it would be hard to do so now as many key people may have died in the intervening period. Would it be appropriate to hold a company liable for something that happened so long ago? For example, if evidence could be found to show that the primary causes of the accident were a failure of the pressure relief door, some might suggest that this were due to maintenance problems. But those responsible for maintenance at JAL then would have long since retired. JAL itself is a different organization now to what it was then.

Finding another cause to the crash may have another consequence. Chapter 5 discussed the compensation paid to *izoku* following JL123, which was done on the basis of the conclusion of the original investigation, although it would appear that JAL and Boeing had agreed the ratio each would pay in advance of the report's publication. If a different cause were found, would this mean that either of these companies would want to find a means to alter the amounts? Although some have suggested to me that JAL may have got a discount on further Boeing purchases after the crash and due to previous orders, others have suggested to me that JAL paid more for some of its subsequent Boeing purchases than was perhaps necessary.[27] So rather than Boeing continuing to show remorse for the JL123 crash and being thankful for JAL's continued custom, apparently JAL was effectively giving Boeing additional money. Was this merely due to poor financial controls and management in JAL at the time? Or was this to compensate Boeing for taking the public blame for the crash, using the maintenance error as a means to show that it was a one-off problem rather than there being a structural problem with the plane, and ensuring that JAL's reputation was not damaged any more than necessary and as much as it would have been the cause been pin-pointed to a maintenance deficiency within JAL?

Although one can continue to speculate about the cause of the crash, in many respects the issue as to who was to blame is secondary. Many *izoku* certainly did not appear to think like this in the early years after the crash, it does appear to be a position which many have moved to. Indeed, one of the requests made by the 8/12 Renrakukai to Transport Minister Maehara in 2010, after he announced that he would attend the twenty-fifth anniversary events at Irei-no-Sono and also visit Osutaka-no-One that day, was that Japan should introduce an investigation system that focuses on discovering the cause of the crash rather than seeking to find who is to blame and whether a crime has been committed. In a speech at Irei-no-Sono, Maehara announced that the government would look into doing this. Additionally, he said that the government would respond to another of the requests that support given to *izoku* after such tragedies be improved. When I first heard that Maehara would be attending the anniversary I was concerned that this could be a *tatemae* exercise, and although his visit was not without issues,[b] the indications are that it really could be the start of a positive process. Maehara revealed a personal interest and concern about JL123 when he commented to me that after seeing the site, he was sad to think about how there could have been more surviviors had the SAR been better. Whilst a twenty-fifth anniversary (*shihanseiki* in Japanese) is often seen as a time for closure in Japan, it would appear that 12 August 2010 was a new dawn in the story of JL123 and not an ending.

Disaster response

While reading many books and having discussions with people about the cause of the crash has been extremely interesting, at one level it has also been a distraction from the primary aim of this study. Although I hope that this study can help the calls for a reinvestigation into the cause of the crash, this study is also an academic study of responses to the JL123 crash, particularly Japanese responses, although it has also considered other responses from around the world. Whilst I have ended up discussing the cause of the crash and whether a reinvestigation may suggest that a crime had been committed, a starting point for this study was the belief that it was some of the responses to the crash which were 'criminal', in terms of it being 'senseless'. When the *Titanic* hit the iceberg its demise was inevitable and all that remained was the question as to how many would survive, which was particularly an issue due to the insufficient provision of lifeboats. The answer for each individual was not determined at the point of the collision, but in many cases until after the ship sank. Similarly, JL123's demise was effectively sealed at 18:24 but the number of survivors was yet to be determined, and would not be finalized until after the plane crashed 32 minutes later. That the SAR was tardy and that foreign help may have been rejected were other factors that decided the final death toll of JL123. These points were what I knew of the crash based on the English-language materials I had come across. But how could this have happened?

Whilst Chapter 8 has suggested that there are 'alternative theories' about the cause of the JL123 crash and that these may also relate to why the SAR activities were delayed, I am yet to be convinced by these and suspect that much of the delay was merely down to a systemic failure. Whilst this failure could in part be blamed on the problems associated with having to deal with something of this scale, surely the state should have mechanisms for preparing for such eventualities. Indeed, in a country which regularly experiences natural events which have the potential to become a disaster, one would expect that the system should be adept at handling major, unexpected events. Whether the cause be natural or man-made, the system should be prepared to respond. Yet, despite the natural challenges Japan faces, it seems in 1985 the system was not able to cope.

How could such systemic problems exist? Much of what happens is based on learnt behaviour. This is perhaps one element of 'culture'. It is not unreasonable

Yoshiyuki Yamamoto 山本仁之 *(44)* · *Yukio Yamamoto* 山本幸男 *(48)*

to think that the response to the fire bombing of Tōkyō in March 1945 may well have drawn upon the experiences of the fires in the wake of the Great Kantō Earthquake in 1923, for example. But how could the seeming unwillingness to accept foreign help in the wake of the Great Kantō Earthquake, as discussed in Chapter 4, still apparently be in place in 1985 or even in 1995 in the wake of the Great Hanshin Earthquake? Is the answer to this more to do with more generic cultural issues rather than established practices and responses, many of which may be explicitly written out? Could it really be that the poor response to JL123 was in part due to Japan being a more group-based, consensus-based society, or the poor communication between bureaucratic bodies? Is it that disaster response needs individuals taking command? Is it that the rotation around posts, meaning that many are generalists rather than specialists in Japanese organizations, leads to an overreliance on set procedures and prolonged discussions, whereas a specialist would be able to draw upon their own knowledge and skills to provide a more rapid response? I suspect that there may be elements of truth in each of these, and in the more than 20 years that I have been studying Japan I have found that my initial admiration of the system of job rotation has been replaced by a much greater level of concern about the capabilities of some to do the job to which they are now assigned and the implications that this has.

Whilst there were clearly problems with the response to JL123, sadly, it has taken other disasters for the realization about how poor the system was so that changes could be made. After the poor response to the Great Hanshin Earthquake, something had to be changed. Now, for example, following a major earthquake, the SDF can be dispatched by command of the Prime Minister without having to wait for a request from the local governor.[28] This is a major improvement and its effectiveness has already been seen with the response to earthquakes in Niigata. Yet, there are still concerns. A study of a disaster prevention area in Kawasaki suggested than it may be unstable,[29] holes have been found in disaster readiness plans following the issuing of tsunami warnings after a major earthquake in Chile in 2010,[30] in 2008 the SDF were slow to respond following a collision with a fishing boat,[31] in that same year only 30 per cent of residents actually recognized the alarm when it sounded for an earthquake in Tōhoku,[32] and some local authorities were not even implementing a new warning system due to budgetary problems.[33] Yet the Japanese can respond to disasters, and Japanese teams have regularly been dispatched overseas to help with relief efforts, such as after the earthquake in Haiti in 2010.[34] Remarkably, there was even discussion about the SDF being dispatched to China following the major earthquake in Sichuan in 2008.[35] Some of the responses to the Great East Japan Earthquake and tsunami in 2011 further revealed that there is probably still much more the Japanese government could do to improve responses to disasters, but it is too soon and beyond the scope of this study to analyse this further here.

Of course we need to remember that the JL123 disaster could have been worse. Had the plane crashed in Tōkyō as it attempted to return to Haneda, for example, then many more could have died. As discussed in Chapter 4, perhaps SDF jets should have been scrambled much quicker and someone in the government

Eiken Yamashiro 山城栄賢 *(35)* · *Michio Yamato* 山登道雄 *(45)* · *Hideki Yamauchi* 山内秀樹 *(41)*

should have been on standby to give the order to shoot down JL123 if there was a risk that the plane could come down in a heavily populated metropolitan area. In a post 9/11 world, although no country has yet needed to make this unpalatable decision, I believe there may be a greater understanding that such a call will one day be needed. That understanding would have been missing in 1985. Although it would be a decision, which if shown to be prudent from a subsequent investigation into the state of the plane and its likelihood to be able to make an emergency landing, would have saved many lives, it would probably be the end of that politician's political career and possibly bring the government down. With JL123 none of this happened. JL123 crashed in mountains and Japan was spared this tragedy being even greater than it already is.

One inescapable part of a major disaster is that there has usually been a loss of life. Although environmental disasters tend to focus upon the loss of non-human lives, most major disasters tend to relate to the loss of human life. It is this element which often makes it a 'disaster' rather than just a natural event or accident. One of the key aims of this study was to try to discover something about not just how the state, organizations and systems respond to a disaster in Japan, but also how individuals respond to them. Chapters 5 and 6 have particularly focused upon the responses of the JL123 *izoku* in 1985 and in subsequent years. A number of features have become clear as a result of this study.

This study has highlighted that it would clearly be folly to suggest that there is a Japanese response to death or disaster. Although this may seem an obvious point, it is not uncommon to find generalizations attempt to suggest that all Japanese, perhaps due to relative homogeneous genetic make-up of the population, think and behave in a common way. In the case of JL123 we have seen how, for example in relation to the continued remembrance of the victims, some do not believe there is any need to visit the crash site, whilst for others this is a very important aspect of the memorialization. Although Chapter 5 highlighted the way in which many Japanese behaved in relation to retrieving the remains of their loved ones, we have to remember that there were many variations. Whilst some *izoku* wanted all the remains of their loved one to be returned, others were less concerned about this. Similarly some *izoku* were not concerned with a proposal for all unclaimed *ihin* to be destroyed, whilst others wanted them to be preserved.[36] The study has also highlighted that whilst many will conduct the traditional, Buddhist-based, rituals following the death of a loved one, there may not be an inherent belief in what they are doing.

Not all *izoku* go to Osutaka-no-One or Ueno-mura, but many do, particularly for the anniversary of the crash. Chapter 6 looked at this process in what I term the Osutaka Pilgrimage. Whilst perhaps not regarded as a pilgrimage by the *izoku* themselves, the Osutaka Pilgrimage contains elements which are common to many other pilgrimages around Japan. Perhaps the most noticeable part, particularly when compared with many other pilgrimages – religious or secular – in other societies is that Japanese pilgrimages tend to be multi-point rather to a single destination. In the case of the Osutaka Pilgrimage this is seen with the *tōrōnagashi* on the evening of 11 August, the visit to Osutaka-no-One and the Irei-no-Sono ceremonies on 12 August. Chapter 6 also pointed out the efforts that have been taken to aid this pilgrimage, such as improving access to Osutaka-no-One and movement around the site itself. These efforts have been effective in ensuring that many *izoku* continue to visit Osutaka-no-One each year. Chapter 6 highlighted that certain years have seen higher than average participation in the pilgrimage and that there appears to be a degree of correlation with significant Buddhist memorial years and other anniversaries.

The nature of JL123 means that those who died can be remembered in a very different way to others who suffer an 'untimely death'. For example, the location in the mountains provides a place of peace for the *izoku*, in a way that could not have happened had the crash happened at an airport. That the plane split up on impact scattering the crew and passengers across the mountainside allowed for individual *bohyō* rather than there being solely a collective memorial. That they died in a crash means there is a specific time when they can be remembered. This provides a focus which does not exist for those who have lost loved ones to cancer, for example. Whilst those who have lost loved ones to such an illness can come together in support groups due to their commonality in the cause of death of their loved one, they will have their own dates which are significant to them. JL123 is unusual and special as it has elements which are both collective and personal in a combination which allows for both group and individual activities.

Traffic accidents or other disasters are also unusual, as was discussed in Chapter 5, as there is often an expectation that there will be some form of compensation. This is separate to the insurance payment which a family may receive if their loved one was covered by such a policy. Why is it that certain types of accident or events lead to these financial payments? Not all, even in the case of JL123, expected it. But if there is to be a formal payment, as opposed to one based on donations from the public, then a system of calculation is needed. The case of JL123 revealed the problems with such a system, with some feeling that consideration needed to be given to the probable mental pain that the victims suffered during the final 32 minutes of the flight, whereas others did not feel this way. Many were also unhappy with the valuation put on their loved one's life, and in some cases even tried turning to the American legal system to get more money from Boeing. But why do some expect payment for families of victims in say a plane crash over and above insurance payments, but not those who have lost a loved one due to an illness such as cancer?

Narratives on JL123 and continued interest

It is not just *izoku* who are visiting Osutaka-no-One. Whilst others connected to the crash in some way may visit on the anniversary, there are many more with no connection to the crash also visiting the site. These are what some may refer to as 'dark tourists'. However, in Chapter 8 I argued that rather than the fact that people died in the crash being a motivating factor for many of these visitors, their inherent interest in JL123, or in some cases their love of the singer Kyū Sakamoto, is probably more significant. Of course there would be no need to visit the site had there not been the crash and deaths, but I think the idea that 'dark tourists' are in any way different to 'normal' tourists, as appears to be implied through the categorization of them into a different group, is misplaced. Indeed, perhaps even the term 'tourist' is misplaced. Are these visitors not also pilgrims? When I read accounts of how pilgrims travel due to a feel that they need to go to the place,[37] that they revisit in their mind many times after a visit[38] (and these days there may, as with JL123, even be videos, some on YouTube, which can aid with this), that they gain a desire to go again,[39] and that one needs time away from the normal busy life to 'focus on themselves',[40] I begin to realize how even my trips to Ueno-mura have moved beyond being merely an academic exercise to being an annual pilgrimage. Other than when we stop to reflect when a family member or friend dies, particularly if it is 'an interrupted life', we seem too busy to think about our own lives and priorities. For me, going to Osutaka-no-One, beyond the academic reasons for being there, has been a regular opportunity to do this away from the distractions of the normal workplace since my phone will not work there. I am perhaps not alone in this respect. Although the number of non-*izoku* who visit Osutaka-no-One more than once may be relatively few, I predict that the number of visitors to Osutaka-no-One will increase, not only due to the increased awareness of the crash due to films and the like, but, as with the visits to World War I graves,[41] as time goes by, more and more people find that they have relatives who died there and so may want to connect with the site.

But why does JL123 remain in the public consciousness, not just in Japan but around the world? Chapter 8 has highlighted that it appears to be a remarkable cocktail of factors coming together that helps to ensure that interest in JL123 is not waning. For example, the crash happened at a time of year when Japan remembers its dead; it is also comes just three days after the anniversary of the bombing of Nagasaki, itself three days after the anniversary of the bombing of Hiroshima; the scale of the crash, for a single plane, was unprecedented; remarkably there were survivors; and the crash involved Japan's national flag carrier, JAL.

Chapter 5 highlighted another of the aspects for which this crash is most well-known; that some passengers wrote *isho*. The actual number, at least those which were found, was small in number, but such writings are unusual and so they provide a rare glimpse into the mind of those who are coming to terms with the fact that they may be about to suffer an 'untimely death'. But, as so few were retrieved, one has to be careful about looking for a common Japanese element to these *isho*. After all it would appear that a typical response was *not* to write an *isho*. But Chapter 5 has highlighted certain points which are of interest about the contents of the *isho* which were retrieved. Having studied JL123 for many years and having spent many long hours on flights between the UK and Japan I have often wondered what I would write should my plane experience trouble. Would I even write an *isho*? If so, would I be doing it due to being influenced by my study of JL123? The calmness of my actual situation when I ponder these thoughts means that any response would probably be somewhat different to if I were to be confronted by such a situation.

Other than visits to Ueno-mura, how does this interest in JL123 manifest itself and what sustains it? Chapter 7 pointed out how the media interest in JL123 has continued, covering of the anniversary events – the Osutaka Pilgrimage – in great detail. This coverage may be much greater than that given to other transportation accidents and so in this respect one has to be careful in drawing too many conclusions about how Japanese respond to disasters through looking at JL123. JL123 clearly has certain elements which are typical, but it is a special crash. That this is the case is highlighted at the anniversary events when other tragedies are also remembered, and *izoku* who lost loved ones in these may also attend the Osutaka Pilgrimage. In part it may be that the media and its consumers do not have the appetite for the reporting of such events throughout the year. However, it would also appear that the scale and symbolic significance of JL123 has meant that it has gained an extra responsibility as the representative for all multiple death accidents and the continued efforts to improve safety. Consequently, the media will continue to report on the activities of JL123 *izoku*, particularly the 8/12 Renrakukai, throughout the year rather than limiting their interest to around the time of the anniversary.

But it is not just the media who are writing about JL123. Ever since 1985, books about JL123 have been written. Whilst the majority of these are concerned with the cause of the crash, or have been written by the *izoku*, there have also been two novels about the crash. Both of these novels have been dramatized and turned into major films, *Kuraimāzu Hai* (2008) and *Shizumanu Taiyō* (2009), bringing JL123 to another generation of Japanese. Although, as Chapter 8 discussed, the content and nature of the two stories was very different – in particular in relation to the use of pseudonyms and whether the flight was referred to using its real name or not – there can be no doubt that such outputs both reflect a continued interest in the crash and that it is likely to further increase interest in the crash. In this respect, JL123 is becoming a part of Japan's 'social memory', just as the sinking of the *Titanic* is part of the UK's.

The timing of *Shizumanu Taiyō*'s release may not have been without its significance in other respects. The film, which focuses as much up the alleged corruption within National Airlines, modelled upon JAL, as much as the 'Osutaka-yama crash', came at a time when stories were beginning to circulate about JAL's own financial problems. Interestingly, while initial media stories about JAL's woes focused purely on the financial aspect, when the publicity for *Shizumanu Taiyō* began, articles referring to safety also began to appear. As this study is about JL123 and not about JAL itself, it is beyond the focus of this study to discuss the variety of reasons for JAL's problems. However, there are a few points worthy of note in relation to JAL's problems in connection to JL123. JAL's domestic revenue was ¥20m less in fiscal 1985 than it had been in fiscal 1984.[42] The impact on JAL of the JL123 crash was clear. I have come across *izoku* who would seemingly not be sorry to see JAL go out of business, saying that the sight of JAL's crane logo, which was dropped in the years following the crash, used to make them feel physically sick. However, I have also come across many more who believe the best legacy to their loved ones would be if JAL went on to become the world's leading airline and be a beacon of high safety standards for the whole industry.[c] The Japanese government also seemed to keen to see JAL survive, with Transport Minister Maehara saying that 'If its flights stop, the Japanese economy as a whole, local economies, and exchange with foreign countries will be seriously hampered'.[43] In early 2010, Amazon's Japanese website began providing details about the sale of the *Shizumanu Taiyō* DVD, but at that time had no image for the cover. For several weeks the image was one provided by a member of the public. This cartoon picture used the original *Shizumanu Taiyō* poster design as a basis, but depicted a bear waving 'bye-bye' with a handkerchief at a setting sun, with JAL's crane logo at its centre. Whilst the sun has yet to set on JAL, the impact of its financial problems have impacted JL123, for it had to borrow ¥12m from the Irei-no-Sono Zaidan to help fund the improvements to the paths at Osutaka-no-One in 2008.[44] In 2011 JAL reintroduced the '*tsuru-maru*' (round crane) logo as symbol of JAL rebirth through its financial rehabilitation.[45] The first plane with the new logo was revealed just a day before JAL used a 747 for the final time.[46]

But JL123 is not merely a domestic story. Like the *Titanic* there is international interest in JL123, but this study has found that the narrative about JL123 in English is somewhat different to the typical Japanese one. The greatest differences are that English language sources have tended to accept the AAIC report's conclusion about the cause of the crash and have tended accept the account a former US serviceman who claimed that a US forces helicopter was at the crash site about two hours after the crash. If one looks at other English language outputs – documentaries, websites, YouTube videos – one can see how the JL123 story has evolved quite differently to those similar outputs in Japanese. It is regrettable that more Japanese with knowledge of this crash have not written in English about JL123 or commented on the English-language materials. Similarly the case of JL123 highlights how the English-speaking world needs to pay more attention to what is written in the native language of cultures it is discussing.

As part of this study I conducted an online survey, as discussed in Chapter 8. Although this survey was limited, the results helped to highlight some of the similarities and differences between Japanese and non-Japanese in relation to JL123. Whilst Chapter 8 highlighted that in relation to why people are interested in JL123 there were not many significant differences between Japanese and non-Japanese, the survey also found that there were significant differences in relation to some other questions. In particular, as Figure 9.2 highlights, the response to whether or not there should be a reinvestigation into the cause of JL123 shows a huge difference. Given the difference in narratives about the crash such a result is perhaps inevitable and is a graphic reminder of how more work needs to be done to alert the Japanese- and English-language worlds to each other's discussions. I hope this book will be one step in this process. However, this is not the only significance of this book. Just as I have said that many books have been written in Japanese in connection to JL123, so I am now adding to the literature on JL123 and this in itself may help to perpetuate the interest, albeit potentially among a different audience. Similarly, as discussed in Chapter 7, although I had gone to Ueno-mura for the first time partly to see how the media reports on the anniversary, I became the focus of media attention and the subsequent interviews were among the stories about JL123. The problems of research influencing and changing the course of the subject being studied should never be overlooked.

Motohiro Yoshikawa 吉川元啓 *(31)* · *Kazuo Yoshimura* 吉村一男 *(43)*

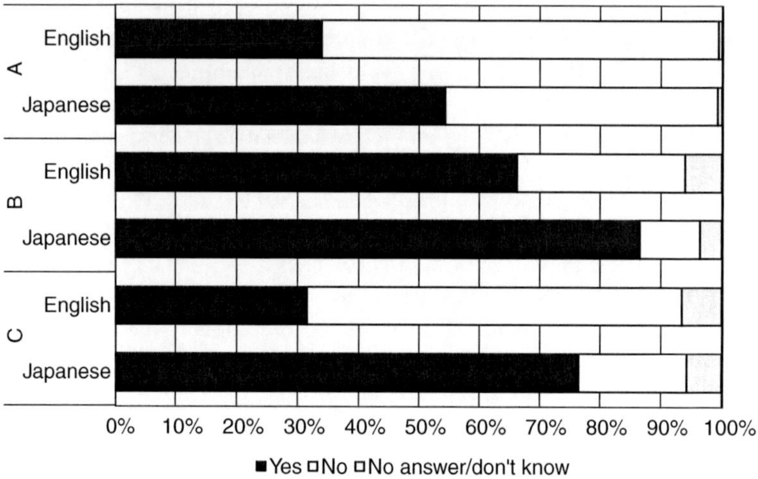

Figure 9.2 Views on JL123 issues

Figure by the author. A: Do you think that the media should publish the passenger list, including details such as their address? B: Do you believe that the Safety Promotion Center is necessary? C: Do you believe that there should be a reinvestigation into the cause of the JL123 accident?

Figure 9.2 also deals with two other issues relating to JL123. The first of these is the media's publication of the passenger list, including personal details such as the home address (as discussed in Chapter 7). Given that it is unusual for detailed information to be ever given by the media in the UK, for example, and that it is only done after the next of kin have been informed, I was somewhat surprised to find that, although most would not want this to happen, the support for it was relatively high. I was also a bit surprised by the Japanese response. In the case of JL123 such information was disclosed by the media, but following changes in the law relating to data protection, it is unlikely that the same degree of information would be given out again. The survey result appears to suggest that there are still many who feel as though a detailed passenger list, with personal information included, has a role to perform. I certainly feel that the provision of a passenger list has its uses, but the nature of the detail and the timing are critical. Initially, more consideration needs to be given to those with next of kin on a flight. These people need to be informed first directly – although this may require them contacting an airline, for example, for confirmation that their loved one was on the flight. They should not be seeing the name on a TV screen. That such a list may contain so little detail that others may mistakenly think they have a loved one on the flight is also a problem. That errors can be made is a further issue that needs to be considered. But in terms of the privacy issues, I do not see why the exact address would be needed – surely the city name would be sufficient. Those who think that they may recognize a name on the list should be able to find another means to confirm whether it is the person or not without the exact address needing to be publicized. But in the longer term, I think a list of names is important for remembrance. That the impact of seeing that children have perished can have a significant emotional impact would suggest that this additional level of detail is also useful in helping achieving the longer term aims of *izoku* groups – that their loves ones are not forgotten and that their deaths are not meaningless, but become a means to promote improvements in transportation safety.

The other issue that Figure 9.2 details is the existence of the Safety Promotion Center in Haneda. The survey shows that there is considerable support for such a centre, particularly among those who completed the Japanese survey. Some of the difference between the levels in positive responses may be due to the phrasing of the original question itself and knowledge of the subject. The survey referred to the Safety Promotion Center as a 'museum' rather than using its name, but I expect that many Japanese respondents would probably be aware that the centre, discussed in Chapter 8, is more than merely a museum. The existence of the centre reveals how JAL's position on JL123 appears to have changed over the years. Based on my discussions with current and former JAL employees, it would seem that JAL almost tried to brush JL123 under the carpet in the early years. Whilst the company may have spoken about the need for safety, it was rarely mentioned in relation to JL123. Those who were in some way involved with JL123, such as having been *sewayaku* looking after *izoku*, were not offered counselling, asked about their experiences, or asked to teach others about those experiences. But things have changed. The Safety Promotion Center is the most obvious physical indication of this. More than 90 per cent employed at JAL have been recruited since JL123. They have no personal connection to the crash. It is important that such employees have the opportunity to go to the centre and even, as many do, go to Osutaka-no-One as it reminds them about the importance of their job and of doing it right.[d] I believe that the centre, and linked activities such as interviewing those who were *sewayaku* so that their experiences can be passed on to others, is an acknowledgement that current employees need to know more about JL123 and that JL123 can be a very effective means to help employees conceptualize what safety is actually about. So how did this change come about? This is not clear. I suspect that the fact that many of those who were *sewayaku* are likely to be in senior management now may be part of the reason. But I believe that it also probably reflects the strength of the 8/12 Renrakukai and the support it has from the media.

However, although JAL's position appears to have changed, one should not assume that there is complete agreement between JAL and the *izoku* on the Center, just as there are likely to be differences of opinions amongst the *izoku*. Both JAL and the 8/12 Renrakukai continue to look for ways to improve the Center and the experience of those who visit there, but I think there needs to be a greater acknowledgement of the level of interest in JL123 and that this interest may be growing. The Center is still relatively small and access is limited, in a location with little obvious connection to the crash itself. Would a larger facility in Ueno-mura be better? Or would the remoteness of this location impact the number of visits? Although some *ihin* have been put on display, there are still many more which have not. Although Mrs Miyajima, head of the 8/12

Renrakukai will speak to JAL employees who visit the Center for training, for example, is there more that could be done? Whilst Mrs Miyajima has become the most well known of the *izoku* and her nine-year-old son, Ken, has become one of the means by which people understand the enormity of the crash, there are many more stories which should be heard. These also need to be recorded so that they are available for generations to come. More care needs to be taken in looking after all of these precious items. They are valuable parts of Japan's social history. Seeing and touching Hirotsugu Kawaguchi's diary which contains his *isho* was very special, but until we spoke about it, even his family were perhaps not fully aware of this item's significance beyond their own feelings. This is not about financial worth – as the case of the *Titanic* has shown that many of these items are becoming almost priceless at auctions – it is about ensuring that the items are preserved for future generations to see so that the lessons about JL123 are not forgotten.

Similarly, more consideration needs to be given how to enhance the experience of non-*izoku* who visit Osutaka-no-One as well as considering how the site should be maintained in the longer term. Even local maps of Ueno-mura, whilst often showing Osutaka-no-One and Irei-no-Sono on them, do not explain what these places are, and I suspect that particularly the name of the latter may have little meaning for those who do not know much about JL123. Walking around Osutaka-no-One, seeing the *bohyō* and memorials that are discussed in Chapter 6, any visitor will surely be impressed at what they see. But it is not easy to fully comprehend the site and what is being seen. I have seen many non-*izoku* visit the site who have no knowledge of the U-shaped cutting, or to have left having not managed to find the *bohyō* for Kyū Sakamoto which they had, at least in part, travelled to see. Many *izoku* no longer visit Osutaka-no-One and so the *bohyō* are deteriorating. The site itself is maintained through the efforts of locals and volunteers who are, or were, at JAL. Although there are many sites like this around the world,[47] I believe that more should be done to improve the visit of non-*izoku* and help those who maintain the site. Should *bohyō* really be left to deteriorate over time? If they are to be left, in another 25 or 50 years will sufficient *izoku* be visiting Osutaka-no-One to maintain the wooden *bohyō*? Or will the site become overgrown and in a state of disrepair like so many graveyards around Japan? Perhaps a group of non-*izoku* who are interested in JL123 would work to ensure that this did not happen. But surely it would be better for the *izoku* themselves to take control of the situation and set out how they would like the site to be maintained and consider how the experience of non-*izoku* visiting the site could be improved in relation to this. Perhaps the 8/12 Renrakukai should even consider recruiting non-*izoku* members to its association, or a new linked association to give direction to this. Based on Walters' observations about the experiences of school children visiting war graves and their understanding being improved due to the connection with visiting the grave of someone who was from their own school,[48] it would be good to see a computer at the museum at Irei-no-Sono which could provide a suggestion about which *bohyō* a non-*izoku* should visit based on their name, gender, age or regional background, for example. The computer could provide more details about the passenger to enhance the connection. Perhaps then there could also be suggestions about what to do at that *bohyō* – clearing away weeds, rubbish or otherwise helping to maintain the site.

The focus of this study has primarily been on the memorialization of those who died in the crash, but I would like to briefly mention the other victims of the crash – the four survivors. All four lost family members or a friend in the crash and all four suffered physical injuries, and, I am sure, mental trauma subsequent to the crash. Based on the searches which ended up accessing my web pages, as discussed in Chapter 8, what has happened to the survivors since 1985 appears to be of much interest. But, as I stated in Chapter 1, I made no attempt to try to contact the survivors myself. The survivors themselves keep a low profile and rarely make any public statements or participate in interviews relating to crash. Only ten years after the crash an article made it clear that Yumi Ochiai, who now had a child, no longer wanted to talk about the crash.[49] Hiroko Yoshizaki spoke to the media around the time of both the fifteenth and twentieth anniversaries and indicated that both she and her daughter had recovered from the crash and were doing well.[50] As for Keiko Kawakami, the iconic image of whom being lifted from the crash site, as well as her subsequent press appearances at the hospital, may have helped to place JL123 in Japan's social memory, she went on to become a nurse, like her mother, and helped at the Great Hanshin Earthquake.[51] In 2002 she got married and had a son in 2004.[52]

In addition to the victims on JL123 and their families, we should not forget those who tried to help them, as discussed in Chapters 5 and 6, and how much these people have also suffered. Those who helped in 1985 in Fujioka are also keen that what happened is not forgotten.[53] There are ways in which the legacy of what happened in response to JL123 will never be forgotten. For example, the daughter of one the dentists who worked on the identification process, and whose enduring memory of that summer was the sound of the helicopters landing and taking off, has gone on to become a dentist herself and helped to develop a system which will radically improve the speed and accuracy of dental identification.[54] Two brothers who are both dentists and JL123 *izoku*, and who helped at Fujioka, also helped with the identification process after both the Great Hanshin Earthquake and the Fukuchiyama Line derailment, and have helped to establish a network for dentists who can provide assistance after disasters and are now pressing for improvements to be made to the process, such as through the standardization of forms not only across Japan but internationally.[55] There were also others – such as SDF soldiers, the police and journalists – who suffered post-traumatic stress disorder as a result of what they saw at Osutaka-no-One in particular. It is still a concern that Japan has not fully developed a proper system for counselling those who need support after such experiences, but one that the government at last appears to be starting to address is due, in large part, to the efforts of the 8/12 Renrakukai and its charismatic, determined leader Kuniko Miyajima who has become a key figure within the NPO movement in Japan as a whole.

Mitsuyoshi Yoshizaki 吉崎充芳 (9) · ***Mikiko Yoshizaki 吉崎美紀子 (8)***

Conclusions

As I mentioned in Chapter 1, I had already started working on another project before a series of coincidences and other factors came together and led to me starting this study. When I started the research I never imagined that it would be as personally fulfilling as it has been academically. It has not always been easy research. The level of detail in Japanese materials about the last 32 minutes of JL123, the nature of the crash, the details about the bodies and their identification was much more graphic than I would have expected in similar books written in English. During the first months of doing this research I had many nightmares in which I dreamt I was on JL123 or a similar plane, or at the JL123 crash site soon after the crash. For me they were just dreams. But for 524 people it really happened. Every time I read an account of the crash, a part of me hoped that the result would be different. Of course it never was. Just as knowing exactly what happened to JL123 will never bring the 521 lives back,[e] history cannot be undone. It is said that 'hindsight is a wonderful thing', but I would have to disagree with this. For the *izoku* who now have the 'benefit of hindsight', and assuming that the conclusions of the official report are accurate, they know how apparently a few relatively simple design alterations could have prevented this accident and their loved ones would still be alive. This knowledge is not wonderful, but a source of frustration and pain.

Today I live in Cardiff, where the popular science-fiction programme *Doctor Who* is filmed. I wish that I could get in his TARDIS and travel back to 12 August 1985 to discover what really happened that day. But time travel would bring a dilemma, often a theme of the sci-fi genre, as to what degree intervention and the changing of events is permissible. Could the lessons that have been learnt since JL123 be learnt without the loss of life? Could the good that has been done by the *izoku* and others have been done without such a tragedy? Could anyone justify allowing the 524 to board JL123 knowing the suffering they were about to endure and the pain that those left behind would have to bear? These will probably forever remain hypothetical questions which nobody will be in a position to confront. So we are left to study the event as a part of history and Japan's social memory. But we must never forget that this was a real event and real people died. For their sake, the sake of those who lost loved ones, for those who tried to help and those who continue to help, for the sake of travellers around the world, we must not forget the crew and passengers of JL123.

But what is it we should be remembering? Good history requires a good story. Perhaps JL123 can aid us all in thinking about who we are. It may even aid the Japanese in learning about their identity. Just as how pilgrimages to the Battle of Kosovo[56] or Little Bighorn[57] can be important to Serbs and Americans, so Osutaka-no-One could be for the Japanese. It is a lesson about the modernization of Japan and the pressures of modern life, and how one responds to a 'problem', which itself is one of the definitions of culture. But perhaps the learning of JL123 should not be restricted to those who 'opt in' through visits to Ueno-mura or the Safety Promotion Center, and that something should be included in the education system or at least there should be materials targeted at children. Indeed, following on from the discussion in Chapter 8 about possible motivations of non-*izoku* to Osutaka-no-One, with Lennon and Foley suggesting that the inclusion of events in the education may have been key in the 'securing an interest' in these events,[58] then such books can be essential for ensuring that a proper understanding is gained. The children and grandchildren of JL123 *izoku* have probably had the key elements relating to the crash described to them sympathetically over time, so as to inform them about what happened and why their family does what it does during the Osutaka Pilgrimage, for example. I doubt other children around Japan get such an introduction. I believe that what is needed are three separate books for elementary, lower secondary and upper secondary school children. The style and level of detail for each should be different. The elementary school one, aimed for children about the age that Ken Miyajima was when he died on JL123, should provide some of the key details and lessons to be learnt about the importance of life. The lower secondary one could build upon this and could be used in moral education classes. The upper secondary version can act as a bridge between the other two books and the more detailed specialist books by authors such as Fujita, Iizuka, Yoshioka, Yoneda, the 8/12 Renrakukai, and the novels and their dramatizations. Each would also need to include *manga*, or for there to be separate *manga* versions, given that this is such a popular form of reading, even for educational materials.

In her 2010 book Kuniko Miyajima writes about the *kizuna* (bonds) with the people who have supported the JL123 *izoku*. But the story of JL123 is about the bonds not just between these living people, but also between these people and those who perished in the crash. The Osutaka Pilgrimage is where all of these various bonds can be seen being reinforced every year. But are the rest of us doing enough to reinforce the bonds that matter? Japan now has 15 national holidays. The only months which do not have a national holiday are June and August. Perhaps consideration should be given to 12 August becoming a safety awareness day. Although transport safety is promoted during the first two weeks of April each year, given how even national holidays are part of the memory of a nation,[59] I believe that it would be fitting for this day to be used in this way given that it also a time when many Japanese are going to be travelling for Obon. The slogan 'no more Hiroshimas' can also mean 'no more Okinawas' or 'no more Chernobyls'.[60] Similarly 'no more JL123s' means 'no more transportation accidents'. But it is more than that. JL123 is not just about a plane crash and death. It is about the importance of life and who is precious to you. It is about the importance of bonds. It is about the importance of the bonds between the living and the dead, and also between the living and the living.

Akihisa Yukawa 湯川昭久 (56) · Yukiyo Zako 雑古祐紀代 (19)

Appendix 1

Boeing 747-SR46 techincal information

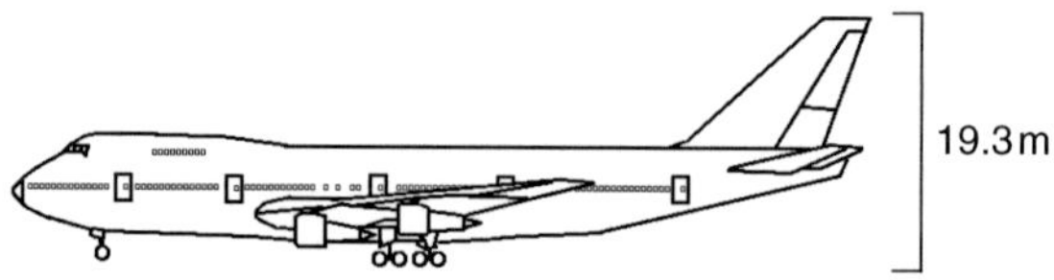

Figure A1.1 Side view of Boeing 747

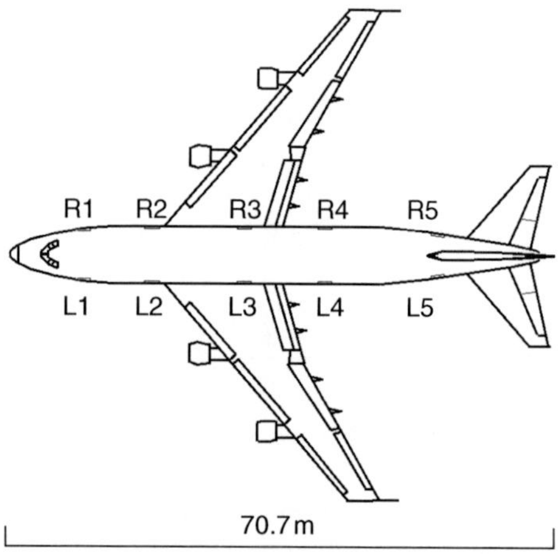

Figure A1.2 Overhead view of Boeing 747 with exit doors labelled

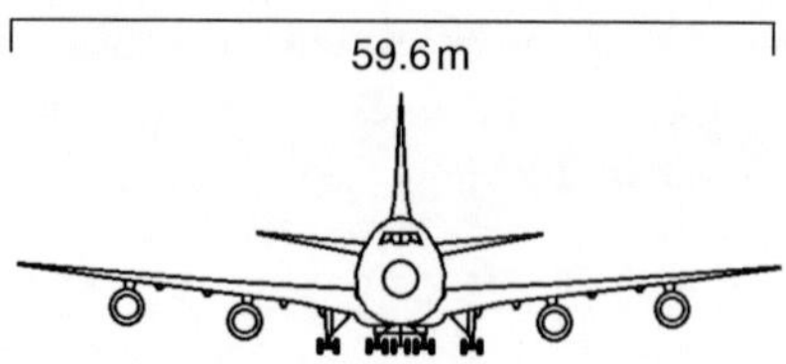

Figure A1.3 Front view of Boeing 747

Figures by the author.

Technical data

Table A1.1

	747-100SR	*747-400*	*777-200*	*777-300ER*
Engine power	4 × 20,930 kg (83,720 kg)	4 × 25,930 kg (103,720 kg)	2 × 33,790 kg (67,580 kg)	2 × 52,160 kg (104,320 kg)
Maximum take-off weight	258.6 t	294.6 t	201.8 t	340.2 t
Maximum landing weight	238.8 t	285.8 t	202.0 t	–
Standard number of seats	530	373	380	246–72
Maximum speed	884 km/h	916 km/h	905 km/h	905 km/h
Range	3,000 km	12,300 km	4,190 km	12,600 km
Required take-off distance	1,680 m	3,250 m	1,820 m	3,430 m
Required landing distance	2,010 m	1,790 m	1,550 m	1,770 m

The above table compares the 747-100SR, similar to the type of plane which was involved in the JL123 crash, a 747-400, which JAL used on many of its international routes, a 777-200, which JAL uses on many domestic routes, and a 777-300ER, which JAL now uses on many international routes. The 747-400 is 5.3 m wider than the 747-100SR and 0.7 m shorter, but the height is the same. The 777-200 is 60.9 m wide, 63.7 m long and 18.5 m high. The 777-300ER is 3.9 m wider, 10.2 m longer and 0.1 m taller.

Source: Kubo 2009:190–1, 194.

Appendix 2

Visiting Osutaka-no-One, Ueno-mura and the Safety Promotion Center

If starting in Tōkyō, the easiest way to travel to Ueno-mura and Osutaka-no-One is to hire a car starting in Takasaki or Chichibu. I would suggest initially going to Ueno-mura and Irei-no-Sono. By going to Irei-no-sono first it is possible to *see* some of the information in the small museum there, as well as the memorial itself, which could aid in the experience of visiting the crash site later. Also it is advisable to eat in Ueno-mura and buy drinks to keep yourself refreshed while you are at the crash site (note: there are no convenience stores in Ueno-mura, so you may find it better to buy food before getting to Ueno-mura).

It is standard for rent-a-cars to have satellite navigation. Start by inputting the following destination: Gunma-ken, Tano-gun, Ueno-mura, Narahara 2218-2 (群馬県多野郡上野村楢原2218-2; Telephone 0274-59-2111; Fax: 0274-59-2470) (the staff at the rent-a-car office should be able to help with this if you cannot use the system or if you cannot input Japanese). If starting in Takasaki, some systems suggest taking the national routes 462 and 299 (indicated by a white number within a rounded blue triangle). This was the way in which the main traffic would have gone in 1985 and also takes you via Fujioka so it is also possible to *see* some of the relevant sites there also. However, although these roads have improved since 1985, the faster route is via the Jōshin'etsu Expressway (上信越自動車道) towards Nagano (長野), leaving at the Shimonita interchange (下仁田IC). This route uses prefectural road 45 (indicated by a white number within a rounded hexagonal blue sign), which has seen improvements over the years – most significantly with the opening of the 3,323 m Yunosawa tunnel (湯ノ沢トンネル). Whichever route you use, you will eventually come to Ueno-mura (上野村) and there will be signs for Irei-no-sono (慰霊の園), taking you off the main road. The final approach up to Irei-no-Sono takes you up a steep road with the entrance on your right. There is free car parking.

Irei-no-Sono

Details about Irei-no-Sono can be found inside the main text of the book. Looking at the main memorial towers, you will *see* there is a small building to the left. In it there is a small museum relating to JL123. At the back of the building there are toilet facilities. Further round to the left there is another building where one can purchase drinks and have a rest. There are also drink machines outside with another seating area and place to leave messages. Walking around the site there are a variety of monuments in addition to the main towers. Behind the towers you will find a memorial where the names of most of the victims names are written (in Japanese alphabetical order starting from the right, with the names of the foreigners in the left-hand most panel). Mid-way along the panels there is a gap, leading to a crypt within the hillside where the cremated remains which were not identified are contained. Before you leave you might also be interested to visit the small memorial on the hill-side, accessed along the path between the main memorial and the museum, to those from Ueno-mura who died during the war.

Ueno-mura village hall

You may also wish to visit the Ueno-mura village hall and surrounding area – particularly if you are in Ueno-mura on 11 August when the *Tōrōnagashi* is held. To get there from Irei-no-Sono, drive back down the hill, and turn right at the bottom of the hill. Follow this road up the hill until you rejoin the main road. Follow the road for about 2 km, through two tunnels. Then take the second left-hand turn after the second tunnel towards the village hall (上野村役場). There is car parking available by a sports ground if you turn left just before you get to the bridge. The village hall is on the right hand side of the bridge.

Osutaka-no-One

Although you can use the car's satellite navigation system (by finding the point on the map and getting it to give directions), there is little need as it is well sign posted and in many areas the reception is poor or the current road deviates from the route in many databases, so is of little use. You need to start by heading west out of the central part of Ueno-mura along national route 299. You will come to a signpost for Osutaka-no-One (御巣鷹の尾根) with a turning to the left just before a bridge, next to a post office. The road to Osutaka-no-One has been greatly improved over the years, with new tunnels being constructed, in order to aid with the construction of Ueno-dam, which you will pass en-route. The road is very narrow in places and frequently suffers from rock falls and landslides. It is therefore recommended that you check that the road is open before departing. Ueno-mura posts information on its website (www.vill.ueno.gunma.jp/koutu and www.vill.ueno.gunma.jp/section/05keizai/douro.html, only in Japanese) and other sites (such as www.goennet.ne.jp/~hohri/n-index.htm, again only in Japanese) also provide updates. There is a small car park at the end of the road, and from there you can then walk to the crash site.

At the car park there is a small map of the site and walking sticks, made by local school children, are provided for those who need them. From the car park to the focal point of the crash site it is about 800 m and in that distance there is a rise of 182 m. Along the path there are a few benches to rest on. Part way you will also find a memorial which has been made for those who are unable to cope with the final climb to the crash site. Here, there are usually free maps of the crash site to take away. Slightly further up you can access a hut for resting in (overnight stays are also possible) and toilets. Further up you reach a point where the path forks (there is a board with a map of the site here), I would recommend that you follow the main path left to Shōkon-no-Hi (昇魂の碑) before looking around the rest of the site. As you go up the main path there are a couple of alternative paths which have more gradual gradients, but will bring you back to the main path further up. Having looked around Shōkon-no-Hi and some of the other areas, you can go round to Sugeno-sawa, where the rear section of the plane eventually rested. You will now be near the bottom of the site and the path will bring you back to the point where you took the left fork (now going straight-on) to Shōkon-no-Hi.

The focal point of the crash site is built upon where one of the helipads was constructed in the wake of the crash. As well as the main memorial stone, Shōkon-no-Hi, there is also a bell and places to leave messages (paper and pens can be found in a box near the bell). If you look to the ridge behind you on your left hand side you should still be able to make out the gauge in the ridge made by the right wing of the plane. Following the path to the left of the bell and up the hillside you will find a number of other memorials – a statue of Kanon, a hut containing various mementos and messages left by *izoku* and visitors and memorial stones detailing the time and date of the crash and the names of the victims. Continuing up this path you will come to the *bohyō* for the captain, co-pilot and flight engineer before coming to the 'Rock X' (*batsu-iwa*), where some of the plane impacted. This effectively marks the highest point of the crash site, so you should now retrace your steps back down to one of the other junctions. If you follow the signs to 4A, 4B, etc. you will be able to get to go round and down to Sugeno-sawa. Due to the scale and topography of the site, even with a map, it can take some time to get used to its layout and I would recommend taking some time to walk around and *see* the variety of *bohyō* designs. The various areas are clearly labelled and so finding the exit to the site should not be a problem when you need to leave. For those wanting to *see* Kyū Sakamoto's *bohyō*, you will need to go to area 7D.

Please remember to respect the site and not to tamper with anything you *see* there. As the car park is very small, and to ensure that the *izoku* are able to visit without problems, non-*izoku* are requested not visit the site before 14:00 on 12 August. Also note that the crash site is inaccessible from around 15 November until 28 April.

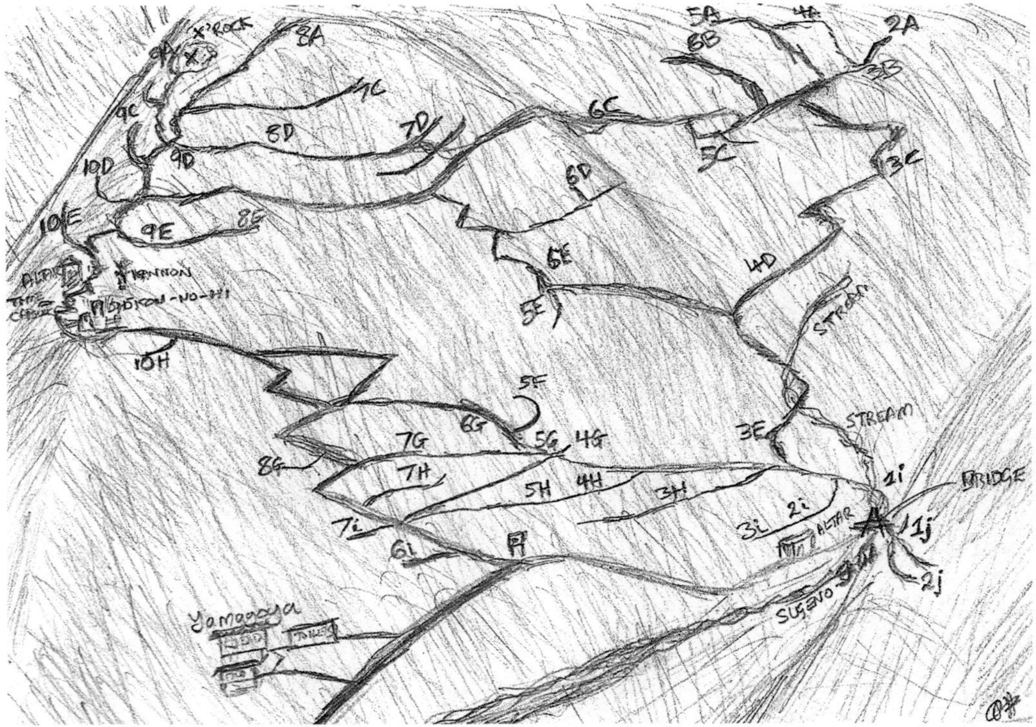

Figure A2.1 Map of Osutaka-no-One

Figure by author. The letters on the map refer to the grid references where remains were found (*see* Figure 5.10)

The JAL Safety Promotion Center

The JAL Safety Promotion Center (*Anzen Keihatsu Sentā*) is located in Tōkyō near Haneda Airport. Although it is possible for the public to visit, visitors have to contact the Center in advance. It is accessed via Seibijō station on the Tōkyō Monorail. Exit the station and go down the stairs to street level, and you should then walk along the pavement directly in front and turn right at the next junction (there is a petrol station on this corner). Continue along the road for about another 200 m and the building in which the Center is located is on your right.

Address: Airport Facilities Co. Ltd, No. 2 Building, 1-7-1 Haneda Aiport, Ōta-ku, Tōkyō 144-0041
Telephone: 03-3747-3391
Fax: 03-3747-4493
Web: www.jal.com/en/safety/center/

Glossary and abbreviations

45/47 System *Yongo-Yonnanna Taisei.* Named using the Japanese system of counting years, it was the policy aimed to control the aviation market in Japan from the early 1970s.

AAIC Aircraft Accident Investigation Commission

Amakudari Literally 'decent from heaven'. Process by which bureaucrats, normally, move into private companies, often related to their previous work upon 'retirement'.

ACC Area Control Centre, a form of air traffic control.

ANA All Nippon Airways.

APC Approach control.

ATC Air traffic control.

Black box Popular term for the flight data recorder and cockpit voice recorder.

Bohyō A marker post for a grave or where a person died (the *kanji* can also read *hakajirushi*).

CVR Cockpit voice recorder (often referred to as the black box).

FAA Federal Aviation Administration of the USA.

FDR Flight data recorder (often referred to as the black box).

Hakajirushi *See Bohyō.*

Hōmyō Buddhist posthumous name (alternative term: *kaimyō*).

Honne The 'real intention' (as opposed to the 'professed intention' (*tatemae*)).

Isho A last note or will.

Izoku Bereaved family.

JAL Japan Air Lines until 1989 when it became Japan Airlines. In 2004 it became Japan Airlines International.

JAS Japan Air System. Originally known as TDA. Became a part of the JAL group in 2002 and effectively ceased to exist in 2004.

Kaimyō Buddhist posthumous name (alternative term: *hōmyō*).

Kanji Chinese characters used in the Japanese language.

Kansai The region of Japan that includes the prefectures of Mie, Shiga, Kyōto, Ōsaka, Hyōgo, Nara and Wakayama.

Kantō	The region of Japan that includes the prefectures of Ibaraki, Tochigi, Gumma, Saitama, Chiba, Tōkyō and Kanagawa.
Kisha Kurabu	Press clubs.
LDP	Liberal Democratic Party.
Meishi	Business card.
NTSB	National Transportation Safety Board of the USA.
Onsen	Hot springs – usually for bathing in, although some are just for observing due to their intense heat.
Ryokan	Traditional accommodation for travellers (often includes dinner and breakfast).
SAR	Search and rescue.
SDF	Self-Defense Forces (*Jieitai*).
Sewayaku	Member of a 'care team' looking after an *izoku*.
Shūkanshi	Weekly magazines.
Sotoba	A post with a Buddhist inscription on. Usually left by a grave.
Tanshinfunin	People who work in a city away from their family.
Tanzaku	Piece of paper for writing messages to usually be left at a shrine or temple.
Tatemae	The 'professed intention' (as opposed to the 'real intention' (*honne*)).
TDA	Tōa Domestic Airlines, formed by the merging of Tōa Airways and Japan Domestic Airlines in 1971. Renamed Japan Air System (JAS) in 1988.
Yakuza	Japanese organized criminal.
Zoku	Group of people with a common interest.
Zokumyō	Name of an individual whilst they were living as opposed to their Buddhist posthumous name (*kaimyō* or *hōmyō*).

The following provide the basis for common conversions:

1 kilometre (km)	=	0.62 miles (1 mile = 1.613 km)
1 metre (m)	=	3.28 feet or 1.09 yards (1 foot = 0.305 m)
1 centimetre (cm)	=	0.39 inches (1 inch = 2.56 cm)
1 metric ton (t)	=	0.98 imperial ton or 1.10 US ton
1 kilogram (kg)	=	2.21 pounds (lb)
1 kilowatt (kW)	=	1.34 horsepower (hp)
1 Nautical Mile (NM)	=	1.852 km
1 knot (kt)	=	1.852 km/h
1 Hectopascal (hPa)	=	100 pascal (pa) = 1 millibar

Notes

1 Introduction

a This text is a translation of the first verse of the song *Sen-no-Kaze-ni Natte* by Man Arai. An English version (known as *Don't stand at my grave and weep*) is generally credited to Mary Elizabeth Frye (1905–2004). However, question marks remain over the origin. This is partly as slightly differing versions exist. Man Arai translated it into Japanese in 2001 and put it to music, and an illustrated book of the poem was published in 2004 (Arai 2004). Arai suggests the origin may date back to a Navaho story and gives no mention to Frye, putting the author as 'unknown'. *Sen-no-Kaze-ni Natte* has become a popular tune in Japan, being covered by a number of different artists. It is now sung at events which commemorate the death of loved ones, such as JL123. I first became aware of the Japanese poem when I received a copy of Arai's book after a friend, Hiromi Tanabe, died of cancer in 2005 aged 30.

b I have also had the nickname *jishin-otoko* whilst in Japan – a name given to those who tend to experience a lot of earthquakes, perhaps with the suggestion that they are in some way a causal factor.

c These are also preferable when taking into account the pressures of work and family commitments. Both of these provide restrictions which few academics seem to discuss in detail when covering the restrictions upon their research.

2 Putting JL123 into context

a The *genkan* is an area where shoes are left before going up the step into the main part of the house. Many consider it to be outside the house, although inside the front door of the house, and visitors may open the door and come into the *genkan* to wait for the owner.

b On 30 July 1971 ANA Flight 58 was on a domestic flight from Chitose to Haneda. As it flew over Shizukuishi in Iwate Prefecture it was hit by a Japanese Self-Defense Force F-86 jet. The pilot of the JSDF jet ejected shortly before the collision and survived. All 155 passengers and seven crew on the Boeing 727 died. It became the world's deadliest air disaster. Source: Gero 2006: 100–2.

c After the first derailment of an in-service shinkansen, following an earthquake in 2004, I was interviewed by some Japanese media and asked whether this was the end of the '*shinwa*' of the shinkansen's safety-record.

d The JMS scale is as follows: 0 – *Mukan* – when nothing occurs; 1 – *Bishin* – a faint tremor, which most will not feel; 2 – *Keishin* – a light tremor, which will cause some shaking; 3 – *Jakushin* – a weak tremor, which causes houses to shake; 4 – *Chūshin* – a medium tremor, which would be felt by most people; 5 – *Kyōshin* – a strong tremor, which would cause some damage; 6 – *Resshin* – a violent tremor, which causes people to fall over and will destroy up to about 30 per cent of housing; 7 – *Gekishin* – a severe tremor, which destroys over 30 per cent of housing. Fractions, such as 4.1, are also possible and descriptions such as a 6-*kyō* (upper 6) or 6-*jaku* (lower 6) may also be used. Source: JMA 2010e.

3 From take-off to disaster

a The AAIC report (197b:6) incorrectly identifies the plane as being a 747SR-100.
b There were seven assistant pursers and four stewardesses, but to avoid confusion I will refer to them all as pursers.
c An adult in Japan is legally defined as being 20 years old, although this may change.
d JA8119 had a seating capacity of 528.
e Departure time is defined at the time when the doors are shut and the plane begins to be pushed back from its stand. 'On-time' is defined as being within 10 minutes of the scheduled time.
f The term 'phugoid' was derived by British aerodynamicist F.W. Lanchester who mistakenly based the term on the Greek word *fugare* ('to flee') rather than *volare* ('to fly') (Job 1996:135).
g The English translation suggests that the Captain said 'Doesn't work. Gear doesn't go down', but this does not fit with the original Japanese – '*Dasenai. Gya orinai'* (AAIC 1987b:Attachment 6, AAIC 1987a:Attachment 6).
h In reality the tree was a Japanese red pine (*Pinus densiflora*) (Kawamura 2004:233).

4 Response and responsibility

a The official report notes the GPS co-ordinates as being 35°59'54N, 138°41'49E (AAIC 1987b:8), which would appear to be halfway between the 'U-shaped cutting' and the main crash site.
b Although forest fires, when compared to places such as Australia, California, Spain and southern France are relatively rare in Japan, they remain a threat in the area and signs around the crash site warn visitors of this. There has been a fire at the crash site probably started by an incense stick and it is the scars of this fire, rather the one after the plane crash, which can be still seen now (interview with Ueno-mura staff).
c Knowledge that planes effectively disappear after some impacts may be more widespread in the wake of *United 93* (2006), based on events about one of the planes hijacked on 9/11, which emphasized that the hole in the ground created by the plane seemed too small.
d An infamous example of this kind of slip is relates to the loss of the *Orbiter* spacecraft on Mars in 1999 when it was realized that NASA was sending commands using imperial measurements when they should have first been converted into metric (BBC 1999b).
e There are a number of such disputes around Japan (*AERA* 2008-09-15)
f *see* cases discussed in BBC 2010.
g There were suspicions, later proven correct, that the earlier 747, Air India Flight AI181, loss had been due to terrorism (Gero 2006:189191).
h In many countries the standard is 10,000 feet, but in Japan it is 13,000 so as to remain above the height of Mt Fuji – 12,500 feet (3,776 m) (Tanabe interview).
i For example a section of text about the empennage is different (AAIC 1987a:10; AAIC1987b:10) and the Japanese version also makes it clear that all doors, some damaged, were recovered at the site, whereas the English version does not (AAIC 1987a:11; AAIC 1987b:11).
j For example, poor construction discovered as a result of the Great Hanshin Earthquake in 1995, the collapse of parts of tunnels on the San'yō Shinkansen in 1999, the 'Structural Calculation Forgery Problem' in 2005, and the global recall of Toyota cars in 2010.

5 Counting the cost

a I have left the original word '*kami-sama*' (deities) as it is ambiguous about whether this would be a request to the Shintō (or Buddhist) deities or a Christian God.
b The original Japanese uses the word '*mama*' ('mother'), a common way for Japanese to address a wife.
c A hotel where couples can go for an hour or longer primarily for sex in surroundings which may cater to certain fantasies and which provide greater privacy than may be found in a traditional Japanese home.
d According to one documentary and the testimony of one of the American team who went to the crash site, flowers were coming down on them whilst they were at the site (Cineflix 2005).
e Although associated with Asian cultures – perhaps due to their use in some Indian and Chinese restaurants, for example – the practice may have originated in Britain. Sheffield University professor Graham Healey has pointed out that Takayoshi Kido, one of the most important leaders of the Meiji Period, spent about four months in the UK in 1872 and his diary entry for 13 September refers to the use of *oshibori* (although he referred to it as a '*tenugui*' ('hand towel')) at a banquet put on at Mercers' Hall. Healey points out that for Kido, the cloths were 'unfamiliar enough for him to mention as curiosities'. As the use of these cloths became more widespread in Japan, the name probably changed to *oshibori* to help distinguish it from other '*tenugui*' (Healey email 2007-07-11).
f *Gannen* is the term usually used to refer to the first year of an Imperial reign. Since the Meiji Period, when an Emperor dies and another Emperor takes to the throne a new name is given to the era, starting from 1 January of that year.
g Kiyoshi Ōkubo abducted, raped and murdered eight women in Gunma in 1971 and was hanged in 1976 (Schreiber 1996:178–97).
h It is relatively unusual for people to visit other people's home in Japan. However, after a death in the family, it is commonplace for friends, family and work colleagues to visit the home of the deceased. In preparation for this, many feel the need to clean the house thoroughly (TBS 2005; *izoku* interviews; Iizuka 2001:141).
i One family continues to use their daughter's, slightly damaged, camera after it was returned to them (TBS 2005).
j PAX=Passengers
k 'High' is an abbreviation for high-heeled shoes.
l For example, the Wikipedia entry on Kyū Sakamoto, which provides no reference for the source of this information. It is probable that many other sites which also include the claim have taken the information from Wikipedia. The original source of the idea is unclear.
m It is ambiguous whether there is a single voice saying this or many.
n Mariko Shirai was a former employee of JAL.
o 17 of the JL123 passengers used JCB Gold Cards to purchase their tickets which meant that there would automatically be a ¥30m payout (anonymous interview).

6 Remembrance and the Osutaka Pilgrimage

a Translation of first two verses of *Miagetegoran Yoru-no Hoshi-o* by Kyū Sakamoto.
b I use '8/12' following the convention of '9/11' and '7/7' that has been used in relation to events in English-speaking countries in subsequent years. Also, the '/' appears on the same key on a standard Japanese keyboard as the '·'
c The name is taken from the colour of the sky around the time of the crash.
d It is important to note that the year of death is counted as being the first year. Consequently, the 'third anniversary' of an individual's death for someone who died in 1985 would be 1987 not 1988.
e A Japanese noun, such as '*senzō*', can be either singular (ancestor) or plural (ancestors), so reference to the '*senzō*' further reflects the ambiguity.

f In the case of JL123, one family of victims left no living relations (JAL interview).
g Indeed, some Facebook groups encouraged British people to enter the name of their favourite football team as their religion in the 2011 census.
h JAL is not alone in missing out certain row numbers. Some Alitalia planes have no row 17 as the number 17, when written in Roman numerals is XVII, an anagram of the word VIXI which in Latin means 'I have lived' and therefore implies 'I am now dead'.
i The six days are *Senshō* (also known as *Senkachi*, *Sakigachi* or *Sakikachi*) – good in the morning, bad in the afternoon; *Tomobiki* – good luck all day other than noon; *Senbu* (also known as *Senpu*, *Senmake*, or *Sakimaki*) – bad luck in the morning, good luck in the afternoon; *Butsumetsu* – unlucky all day (the day Buddha died); *Taian* – the day of great peace and best for ceremonies; *Shakkō* (also known as *shakku*, *jakkō* or *sekiguchi*) – bad luck all day except at noon.
j This belief has also been the basis as to why many Japanese do not mark their bodies, for example getting their ears pierced or having tattoos.
k Sanzu River is the river that the dead spirits are thought to cross on the seventh day after death to the next world.
l Homes will often also have a Shintō *kamidana*.
m Hanshin Tigers did not win the Central Championship in 1995, but Orix BlueWave, based in Kōbe, with the help of Ichirō Suzuki who went on to have great success in the Major League with the Seattle Mariners, won the Pacific League title and were runners up in the Japan Series.
n A professor from Ritsumeikan University and some of his students might have been amongst the first at the site. The details of how they found the crash site and what they did at the crash site are patchy, and there are some inconsistencies with the reports (Tsunoda 1993:293–5). The professor did not want to discuss it when I attempted to contact him about it.
o I use 'Osutaka' rather than 'Osutaka-no-One' as 'Osutaka' has become so synonymous with the crash.
p It should be noted that people will often also visit other sites, including Shintō shrines and other temples not included in the 88 Shingon Sect temples, during the pilgrimage (Reader 1993:110).
q Note that a village hall in Japan is the administrative building where the mayor and other local administrators work and not, as would be the case in the UK, a building designed for community activities.
r JAL had to make special arrangements to ensure that *izoku* would be able to access Osutaka-no-One on 12 August.
s Literally 'a thousand cranes'. These are chains of origami cranes which are thought to bring good luck and ward off illness. However, they have also become connected with the remembrance of the bombing of Hiroshima and have also become a symbol of peace and remembrance.
t 520 cherry trees were also planted at Ishiyama-dera in Kyōto by a member of the 8/12 Renrakukai.
u *Bohyō* are sometimes called *hakajirushi*, an alternative reading for the same *kanji* (Chinese characters) with which the word is written.
v Remember that larger versions can be found at www.hood-online.co.uk/JL123.
w On the front of all three posts posthumous names are used, while the original names appear on the back. Consequently it is hard for those unfamiliar with the site or these individuals to find where the memorial to these three people is.
x Jizō is a bodhisattva who looks after children who die before their parents, of travellers and of fire-fighters, so it is not surprising to *see* such statues at Osutaka-no-One.
y *see* Uenomura no Chishi Shippitsusha 2003:73 for more information about this.
z It is somewhat ironic that *Osutaka-no Sōkatachi* contains a slip of paper referring to an error on one the pages of the book and that the error appears on page 123.

7 The media and JL123

a This quotation is taken from a conversation in the novel *Kuraimāzu Hai* (*see* Chapter 8) (Yokoyama 2006:426).
b One salesmen who visited my flat in Japan suggested that by not taking out a subscription, he would miss out on getting his target so would not get enough money to pay to feed his family.
c As NHK has to raise its income through license fees it is often compared with the BBC. However, with its budget being agreed by the Diet and pressure from politicians to influence programme contents (Altman 1996:170), NHK clearly does not enjoy as much independence as the BBC.
d *Kanji* can have multiple readings in Japanese and for names in particular there can be more than one way for a name to be written, which is one of the reasons why it also took the newspapers so long to prepare the list.
e Although the main article will always be in the top right hand quadrant of a newspaper, there may not always be a headline across the top of the paper. If there is one, it might only be across one-half of the page. If there is no top headline, the side headline, on the right hand side, is the main headline for the paper.
f *Chan* is the suffix often added to girl's names, in particular, in the place of *san*. *Kun* may be added to boy's names in this way.

8 Japan's *Titanic*

a Some journalists remember seeing a number of shooting stars that night as they searched for the crash site (journalist interviews; Asahi Shimbun Shakaibu 2006:131).
b As discussed in the documentary, she went on to become a stewardess for JAL.
c Whilst this was the first time that JL123 featured as the main story, it has appeared in other *manga* in a variety of ways over the years. For example, one *manga* uses imagery which would appear to be based upon the rescue of the four survivors of JL123 (Ōtsuka and Yamazaki 2009:158).
d The official international title for the film *Shizumanu Taiyō* is 'The Unbroken', which has no relationship to that Japanese title. This is not the only case where English and Japanese titles are significantly different. Kyū Sakamoto's *Ue-o mite arukō*, became 'The Sukiyaki song' rather than 'I look up when I walk'. The Oscar winning films *Sen to Chihiro no Kamikakushi* (Sen and Chihiro's Spriting Away) and *Okuribito* ('person who seems someone off') became 'Spirited Away' and 'Departures' respectively. Changes happen also for foreign films when they are shown in Japan. The emphasis of the James Bond film set mostly in Japan shifted from 'You only live twice' to '007 dies twice' (*007-wa Nido Shinu*).
e There is also an English play *Charlie Victor Romeo* which dramatizes various cockpit voice recordings from accidents, including JL123. *see* www.charlievictorromeo.com
f www.hood-online.co.uk/JL123 (and www.jl123.co.uk between August 2007 and November 2010) and cardiff.academia.edu/ChristopherHood.
g There are apparently many *izoku* who feel that this apartment should be maintained as part of a memorial due to its iconic nature (Miyajima interview 2009).
h This sign is already some kilometres along the road to Osutaka-no-One by which time the visitors are likely to have already travelled some distance to Ueno-mura.
i I have come across some such 'dark tourists' who were struggling to find Sakamoto's *bohyō*. In another case, at Irei-no-Sono, of 520 names, only the name Hisashi Ōshima (Sakamoto's real name) appeared to have any resonance with the group, although they mispronounced the *kanji* for 'Hisashi' as 'Kyū'.

9 Conclusions: JL123 present and future

a ARAIC was established in 2001, replacing the AAIC.

b When Minister Maehara was walking up to Shōkon-no-Hi he was walking faster than some *izoku* and one of his staff suggested that the *izoku* should walk faster. Given the insensitivity of such a comment, the *izoku*'s retort of 'Edo *jidai'* ('Edo Period' – i.e. a traditional way of thinking dating back to when *daimyō* and samurai were most important) was very measured.

c I have come across many *izoku* who no longer use JAL, but have also come across others who do not mind. I have even come across one who wanted to become a pilot despite losing his parents and sister in the JL123 crash.

d Similarly, a company, Eikō, has made groups of new employees do the Shikoku *henro* as part of their training following an accident in 1965 when a child was killed by a company truck (Reader 2005:81–2).

e As noted elsewhere in the book, although officially there were 520 victims, one of these also lost her foetus in the crash and so 521 lives were lost in the crash.

References

1 Introduction

1 *See also* Naitō 2005:20.
2 Hood 2010.
3 Hood 2006:5–6, 149–55.
4 Kluckhohn 1949:17.
5 Hood 2006:6.
6 Hood 2001:2 2006:4.
7 Hood 2001:2,5 2006:7–8.
8 Hood 2006:13–14.
9 Reader 2005:225.
10 Hood 2002.
11 Hood 2006:5.
12 Hood 2006:8–9.

2 Putting JL123 into context

1 Miyazaki 2005:318.
2 Asahi Shimbun English Web Edition 2005.
3 Dumbrell 2001:89.
4 Hood 2001:34; 39.
5 Endow 2003:28
6 BBC 2003.
7 Yoshizaki 2005:15.
8 Yoshizaki 2005:11–37.
9 BBC 2001.
10 Yoshizaki 2005:131.
11 Sōmuchō 1987.
12 Sōmuchō 1987.
13 BBC 2005c.
14 Sōmushō 2010a.
15 Schreiber 1996:254–75.
16 Shinjinbutsuōraisha 2010:248; Yoshizaki 2005:96–108.
17 Horsley and Buckley 1990:108.
18 Yoshizaki 2005:104.
19 *The Japan Times* 1985-08-04.
20 MLIT 2010a.
21 Reader 1993:123; Reader 2005:22
22 Oedewald 2009:113.
23 Sōmuchō 1987:300; Sōmushō 2010b:388.
24 Kingston 2004:153.
25 MLIT interview.
26 Hood 2006:7.
27 MLIT 2010b.
28 Sugiura 2001:155.
29 JAL 2002:7.
30 JAL 2002:7.
31 JAL 2002:6,7.
32 JAL 2002:8.
33 JAL 2002:9.
34 JAL 2002:9.
35 JAL 1985b:1.
36 JAL 2002:11.
37 JAL 1985b:1.
38 JAL 1985b:2.
39 JAL 1985b:2.
40 Kubo 2009:181.
41 JAL 1988:3.
42 JAL 1988:3.
43 JAL 1985a:1.
44 JAL 1988:3.
45 JAL 1988:4.
46 JAL 2002:11,13; Cabinet Secretariat 2010.
47 JAL 2002:11,13.
48 JAL interview 2004; Hood 2006:39.
49 JAL 2002:CD-ROM.
50 JAL 2002:CD-ROM.
51 JAL 2002:CD-ROM.
52 JAL 2002:CD-ROM.
53 JAL 2002:11
54 JAL 1988:4.
55 JAL 1988:4.
56 JAL 1988:4.
57 JAL 1988:5.
58 JAL 1988:6.

59 JAL interview 2007.
60 JAL 1988:6.
61 Asahi Shimbun Shakaibu 2006:19.
62 Gero 2006:189–91.
63 Gero 2006: 191–3.
64 Gero 2006: 195–8.
65 Gero 2006: 198–200.
66 Stewart 1988:11.
67 Boeing 2009:18.
68 Barlay 1990:1–2.
69 Cobb and Primo 2003:46.
70 Weir 2000:1.
71 Cobb and Primo 2003:1.
72 ACRO 2010b.
73 Wiegmann and Shappell 2003:116–17.
74 Tanabe interview.
75 ACRO 2010b.
76 Boeing 2009:22.
77 ACRO 2010b.
78 NPA 2003:12; ASN 2010a; ASN 2010b; ASN 2010c; ASN 2010d; ASN 2010e; ASN 2010f; ASN 2010g; ASN 2010l.
79 ASN 2010h.
80 ASN 2010i; JAL interview.
81 ASN 2010j.
82 Haines interview; Tudor email 2010-10-01.
83 ASN 2010k; Saigai Jōhō Sentā and Nichigai Asoshiētsu Henshūbu (eds.) 2007:103–4.
84 ASN 2010m; JAL interview.
85 Levinson and Granot 2002:xi.
86 Cutter 2005:39.
87 Thang 2005:199.
88 Woods and Woods 2007:32.
89 Noguchi and Fujii 2000:52.
90 FDMA 2010.
91 JMA 2010d:20.
92 Teramura interview 2009.
93 Alexander 2005:30.

3 From take-off to disaster

1 Yoshioka 1986:80.
2 Yoshioka 1986:68–80, 131–7.
3 Kōdansha 2010.
4 Weir 2000:114–20.
5 Job 1996:136–51.
6 Kubo 2009:181.
7 Saigai Jōhō Sentā and Nichigai Asoshiētsu Henshūbu (eds) 2007:341.
8 Asahi Shimbun Shakaibu 2006:236.
9 AAIC 1987b:17.
10 AAIC 1987b:6.
11 AAIC 1987b:6.
12 AAIC 1987b:14.
13 AAIC 1987b:14–15.
14 AAIC 1987b:15.
15 Job 1996:136.
16 AAIC 1987b:15–16.
17 AAIC 1987b:15–16.
18 Reader 1993:131.
19 National Defense Counsel for Victims of Karōshi 1990:vi, vii.
20 AAIC 1987b:19; Goo 2010a.
21 Yoshioka 1986:11.
22 AAIC 1987b:6.
23 BBC 2005a.
24 AAIC 1987b:6.
25 AAIC 1987b:6.
26 AAIC 1987a:Attachment 6.
27 AAIC 1987b:6.
28 AAIC 1987a:6.
29 AAIC 1987b:6, 85.
30 Yoshioka 1986:69.
31 Yoneda 2005:199; AAIC 1987b:92.
32 Yoneda 2005:128,144.
33 AAIC 1987a:Attachment 6.
34 Yoneda 2005:46.
35 AAIC 1987a:Attachment 6.
36 AAIC 1987a:Attachment 6.
37 Yoshioka 1986:72.
38 AAIC 1987a:Attachment 6.
39 AAIC 1987a:Attachment 6.
40 AAIC 1987a:Attachment 6.
41 Anderson and Eberhardt 2010:96–7.
42 Anderson and Eberhardt 2010:97.
43 Yoshioka 1986:73, 75.
44 AAIC 1987a:Attachment 6; AAIC 1987a:Attachment 5.
45 AAIC 1987a:Attachment 6.
46 Brus 1999.
47 *The Economist* 2002.
48 Bartlett 2009:26.
49 Bartlett 2009:27.
50 Bartlett 2009:15.
51 Bartlett 2009:27.
52 AAIC 1987a:Attachment 6.
53 Tanabe interview January 2009.
54 AAIC 1987a:Attachment 6.
55 AAIC 1987a:Attachment 6.
56 AAIC 1987a:Attachment 6.
57 AAIC 1987b:115.
58 AAIC 1987a:Attachment 6; AAIC 1987a:Attachment 5.
59 AAIC 1987a:Attachment 6.
60 AAIC 1987a:Attachment 6.

61 AAIC 1987a:Attachment 6.
62 AAIC 1987b:117.
63 AAIC 1987a:Attachment 6.
64 AAIC 1987a:Figure 1.
65 AAIC 1987a:Attachment 6.
66 AAIC 1987a:Figure 1.
67 AAIC 1987a:Attachment 6.
68 AAIC 1987a:Attachment 6.
69 AAIC 1987a:Figure 1; AAIC 1987a:Attachment 5.
70 AAIC 1987a:Attachment 6.
71 AAIC 1987a:Figure 1.
72 AAIC 1987a:Attachment 6.
73 AAIC 1987a:Attachment 6.
74 Yoshioka 1986:74–9.
75 Kawamura 2004:257.
76 AAIC 1987a:Attachment 6; AAIC 1987a:Attachment 5.
77 Yoshioka 1986:79.
78 AAIC 1987b:80.
79 AAIC 1987b:80; Kawamura 2004:235.
80 AAIC 1987b:80.
81 AAIC 1987b:108.

4 Response and responsibility

1 Letter from Lyman Cotten to Elizabeth Henderson Cotten, 1923-09-15 cited in Hammer 2006:230.
2 Asahi Shimbun Shakaibu 2006:18; Yoneda 2005:60.
3 Asahi Shimbun Shakaibu 2006:18; Yoneda 2005:61.
4 Asahi Shimbun Shakaibu 2006:71.
5 Antonucci 1995.
6 AAIC 1987b:23.
7 Antonucci 1995.
8 Antonucci 1995.
9 Antonucci 1995.
10 Kawamura 2004:27.
11 Kawamura 2004:28.
12 Asahi Shimbun Shakaibu 2006:143.
13 Asahi Shimbun Shakaibu 2006:88–9.
14 Yoneda 2005:81.
15 *Izoku* interview.
16 Kawamura 2004:94.
17 Yoneda 2005:65.
18 Kawamura 2004:90–1.
19 Yoneda 2005:83.
20 Uenomura no Chishi Shippitsusha 2003:44,46.
21 Kawamura 2004:2.
22 Ishibuchi 2003:155–6.
23 Interviews with people from Ueno-mura.
24 Email from Prof. Walter Gear (Cardiff University) on 2008-03-31.
25 AAIC 1987b:128.
26 Yoneda 2005:58; Kawamura 2004:174–5.
27 McKay interview.
28 Yoneda 2005:58.
29 Yoneda 2005:58–9.
30 Yoneda 2005:59.
31 TBS 2005.
32 Yoneda 2005:59.
33 Yoneda 2005:59; Fujita 2005b:44.
34 *Izoku* interviews.
35 Asahi Shimbun Shakaibu 2006:198-200; Yoneda 2005: 28–9.
36 Iizuka 2001:233–7.
37 Yoshioka 1986:132–5; Yoneda 2005:43–8.
38 Kawamura 2004:39.
39 Yoneda 2005:75.
40 Yoshihara 1985b:27.
41 AAIC 1987b:129.
42 TBS 2005.
43 Asahi Shimbun Shakaibu 2006:237.
44 *Yomiuri Shimbun* 1985-8-15.
45 Yamazaki interview.
46 AAIC 1987b:97.
47 Yoshioka 1986:71, 73.
48 TBS 2005.
49 TBS 2005.
50 TBS 2005.
51 TBS 2005.
52 TBS 2005.
53 AAIC 1987b:17.
54 Weir email 2010-10-20.
55 Weir email 2010-10-20.
56 Fujita 2005b:133–4.
57 Gero 2006:342–4.
58 *New York Times* 2008, BBC 2008a.
59 Katō 2006a:144–52.
60 BBC 2008b.
61 Spence 2007:320.
62 AAIC 1987b:85.
63 TBS 2005.
64 Yoneda 2005:177.
65 Taylor email 2008-04-03.
66 AAIC 1987b:13.
67 *Izoku* interviews; Fujita 2005b:182–6.
68 *Izoku* interviews.
69 Yoneda 2005:123–68; Fujita 2005b:177–82.

70 Miyajima interview.
71 Fujita 2005b:233.
72 Yoneda 2005:199.
73 Ogawa interview.
74 Yoshioka 1986:69.
75 Anonymous interview.
76 Anonymous interview.
77 AAIC 1987b:1–3.
78 TBS 2005; BBC 1999a.
79 Yanagida interview.
80 Kawamura 2005:223.
81 IFALPA, no date.
82 van Wolferen 1989:320.
83 AAIC 1987b:124.
84 Weir 2000:257.
85 Weir 2000:356–7.
86 Ishihara 1989:39.
87 Ishihara 1989:38–9.
88 Repeta interview.
89 Seeman 1987.
90 Repeta interview.
91 Kingston 2004:43.
92 Fujita 2005b:125; Repeta interview.
93 Fujita 2005b:125–32; TBS 2005.
94 JAL 1985c.
95 Benedict 1946.
96 Reader, Andreasen and Stefánsson 1993:30.
97 Anonymous interview.
98 Reader, Andreasen and Stefánsson 1993:30.
99 Reader, Andreasen and Stefánsson 1993:31.
100 JAL 1985d.
101 *Flight International* 1985-11-16.
102 Haruta and Hallahan 2003:143.
103 *Flight International*, 1985-11-16.
104 AAIC 1987b:60; Tsunoda 1993:291.
105 van Wolferen 1989:319–20.
106 *New York Times* 1988.
107 Kawamura 2004:142; van Wolferen 1989:319.
108 van Wolferen 1989:319.
109 van Wolferen 1989:319.
110 van Wolferen 1989:319.
111 Kawamura 2005:231–3.
112 *The Japan Times* 1985-08-15.
113 Pharr 1986:56.
114 *Asahi Shimbun* 1985-10-31.
115 van Wolferen 1989:320.

5 Counting the cost

1 Iizuka 2001:32-3.
2 Kawamura 2004:193–4; Iizuka 2005:175-7, 206–8.
3 Gunma-ken Sōmubu 1986:20.
4 Iizuka 2005:204.
5 Iizuka 2005:183–4, 203.
6 Iizuka 2005:202.
7 Kawamura 2005:178.
8 Kawamura 2005:176.
9 Kawamura 2005:186.
10 Kawamura 2005:184.
11 *Izoku* interviews.
12 *Izoku* interviews.
13 Asahi Shimbun Shakaibu 2006:127–9.
14 Asahi Shimbun Shakaibu 2006:127.
15 Kadota 2010:131; *izoku* interviews.
16 e.g. Iizuka 2001:90–1.
17 Kadota 2010:134.
18 JAL interview.
19 Kadota 2010:39, 273.
20 Kadota 2010:41, 274.
21 Kadota 2010:274.
22 JAL employee cited in Haruta and Hallahan 2003:137.
23 Fujita 2005b:211–22.
24 Darren McDonald interview.
25 Haines email 2008-06-11.
26 Yoshioka 1986:37, 40.
27 *Izoku* and JAL employee interview.
28 Yoshioka 1986:40.
29 Yoshioka 1986:40.
30 JAL interview.
31 Yoshioka 1986:28.
32 JAL interview; *izoku* interview.
33 Goo 2010b.
34 Mathews interview; Yoshioka 1986:40.
35 E.g, Fuji TV 2005.
36 *Izoku* interviews.
37 JAL staff interviews.
38 Yoshioka 1986:36.
39 Yoshioka 1986:33.
40 Yoshioka 1986:54.
41 *Asahi Shimbun* 1985-09-04.
42 *Asahi Shimbun* 1985-09-04.
43 Waddington 2009.
44 Reader 1993:115; Miyata 1988:97–128.
45 Nakano 2005:1.
46 Chiba 1997:4.
47 Nakano 2005:1.

48 Scott 2008:188.
49 Hui 2005:208.
50 Scott 2008:189.
51 *Asahi Shimbun* 1985-09-04.
52 *Asahi Shimbun* 1985-09-04.
53 *Asahi Shimbun* 1985-09-04.
54 *Asahi Shimbun* 1985-09-04.
55 *Asahi Shimbun* 1985-09-04.
56 Iizuka 2005:204.
57 Iizuka 2001:64–6.
58 Kobayashi 2010.
59 Gunma-ken Sōmubu 1986:26.
60 Iizuka 2005:278.
61 Mathews interview.
62 Kawahara interview.
63 Iizuka 2001:56.
64 Kawakita 1992:38.
65 *Izoku* interviews.
66 Kawakita interview; Kawakita 1992.
67 Kawakita interview; Kawakita 1992.
68 *Yomiuri Shimbun*, 2010-08-17; *The Japan Times*, 2010-03-17.
69 Kawamura 2005:102–3.
70 Iizuka 2001:174.
71 Iizuka 2001:99.
72 Kawamura 2005:68–9.
73 Iizuka 2001:162.
74 Iizuka 2001:118.
75 Yoshioka 1986:206; Aoyama 2010:405.
76 Kawamura 2005:119.
77 Iizuka 2001:174.
78 Kawakita interview; Kawakita 1992:50–3.
79 Levinson and Granot 2002:154–6.
80 Levinson and Granot 2002:165–6.
81 Kosuge interview.
82 *Izoku* interview.
83 *Izoku* interview.
84 Kadota 2010:184.
85 Ferguson 1988:14–15.
86 Walter 1993:79.
87 Kawaguchi interview.
88 Yoshioka 1986:76.
89 TBS 2005.
90 Yoshioka 1986:77.
91 JAL 1985e.
92 JAL 1985e.
93 *Asian Wall Street Journal* 1987-06-27.
94 *Sankei Shimbun* 1985-08-20.
95 Yoshioka 1986:250.
96 JAL 2001.
97 Yoshioka 1986:246.
98 *Izoku* interviews.
99 Anonymous interview 2009.
100 BBC 2009.
101 Haruta and Hallahan 2003:141.
102 8/12 Renrakukai 2006:51.
103 JAL 2001.
104 Susanne Bayly and Diana Yukawa interview.
105 Susanne Bayly and Diana Yukawa interview.

6 Remembrance and the Osutaka Pilgrimage

1 8/12 Renrakukai 2006:22; Miyajima interview.
2 Miyajima interview.
3 Miyajima interview; 8/12 Renrakukai 2006:24.
4 8/12 Renrakukai 2006:22.
5 8/12 Renrakukai 2006.
6 8/12 Renrakukai 2005:4.
7 *Izoku* interview.
8 Miyajima interview.
9 Based on interviews with a number of *izoku* and JAL staff.
10 Based on interviews with a number of *izoku*.
11 Kawakita interview.
12 *Asahi Shimbun* 2003:237.
13 Reader and Tanabe 1998:7.
14 Reader and Tanabe 1998:5.
15 Reader and Tanabe 1998:5.
16 *see* Hood 2012.
17 Reader and Tanabe 1998:8.
18 Reader, Andreasen and Stefánsson 1993:46.
19 Earhart 1974:61.
20 Reader and Tanabe 1998:15; Reader, Andreasen and Stefánsson 1993:53.
21 Miyajima interview.
22 Earhart 1974:61.
23 JAL 1985f.
24 Reader and Tanabe 1998:15.
25 Nakamaki 2005.
26 van Bremen 2005:26.
27 Tanakamaru 2002.
28 van Bremen 2005:26.
29 van Bremen 2005:27.
30 Kluckhohn 1949:17.
31 Hadfield 2001.
32 BBC 2004a, BBC 2004b.
33 Kokubu 2008.
34 Kawamura 2005:242–3.

35 Miyake 1999.
36 Iizuka 2001:192.
37 Yoshioka 1986:217.
38 Bernstein 2006:31,67.
39 Schulz 2009 (follow-up question after presentation).
40 Kawamura 2005:196.
41 Kawamura 2005:202–4.
42 Kawamura 2005:205.
43 Kawamura 2005:205.
44 Kawamura 2005:205–6.
45 Kawamura 2005:240.
46 Kawamura 2005:210–1, 245.
47 Kawamura 2005:217.
48 Suzuki 2012.
49 Reader, Andreasen and Stefánsson 1993:52.
50 Earhart 1974:62.
51 Nakamaki 2005:45.
52 Iwasaka and Toelken 1994.
53 *Yomiuri Shimbun* 2010-05-01.
54 Iizuka 2006:119–20.
55 Iizuka 2006:120–1.
56 Iizuka 2006:99.
57 Kitagawa 1987:127.
58 Reader 1993:5.
59 Walter 1993.
60 Moore 1980.
61 Notoji 1990.
62 King 1993.
63 Bowman 1993:56.
64 Reader 1993:8.
65 Walter 1993.
66 Moore 1980; Notoji 1990.
67 King 1993.
68 West Sellars and Walter 1993:182.
69 Reader 1993:17.
70 Guichard-Anguis 2009:8; Oedewald 2009:113.
71 Reader 2005:10.
72 Turner and Turner 1978:20.
73 Reader 1993:112.
74 Reader 1993:112.
75 Reader 1993:113.
76 Oedewald 2009:113.
77 Reader 1993:108.
78 Reader 1993:231.
79 Reader 2005:108.
80 Reader 2005:280.
81 Uenomura no Chishi Shippitsusha 2003:120–4.
82 Tamaru and Reid 1996:81.
83 Kitagawa 1987:128–9.
84 Reader 2005:49.
85 Uenomura no Chishi Shippitsusha 2003:181.
86 Anonymous interviews.
87 *Yomiuri Shimbun*, 2009-07-26.
88 *Yomiuri Shimbun*, 2006-03-28.
89 Interview, 2009-08-12.
90 *Yomiuri Shimbun*, 2009-07-26.
91 Reader and Tanabe 1998:22.
92 Reader 2005:68.
93 Reader 2005:69.
94 IREI Air Safety 1999.
95 van Bremen 2005:29.
96 Irei-no-Sono staff interview August 2009.
97 Irei-no-Sono staff interview August 2009.
98 Irei-no-Sono staff interview August 2009.
99 Hagio interview August 2009.
100 Chūgoku Shimbunsha 1986:55.
101 Miyajima interview.
102 Nakamaki 2005:55.
103 Interview August 2009.
104 Reader 1993:232.
105 *Mainichi Gurafu*, 1986-08-31.
106 Interviews, January 2009.
107 Mainichi Shimbun 2010-08-03.
108 Reader, Andreasen and Stefánsson 1993:48.
109 Reader and Tanabe 1998:59, 182.
110 Reader and Tanabe 1998:109.
111 Reader interview 2007; Reader and Tanabe 1998:53.
112 Reader 1993:112.
113 JAL email 2007-11-02.
114 JAL interview.
115 Reader 1993:111; Reader 2005:63.
116 Miyajima interview.
117 Reader 2005:73.
118 Kawaguchi interview 2010.
119 Reader 1993:10–11.
120 Reader 1993:11.
121 Reader 2005:192.
122 Aoki 1994:165–70.
123 Reader 1993:237.
124 Interview August 2009.
125 Reader 1993:124-5; Befu 1983:259.
126 Komatsu 2000; Komatsu 2002.
127 *Asahi Shimbun*, 2003-07-31.
128 *Asahi Shimbun*, 2005-07-31.
129 Reader 1993:234.
130 Reader 1993:235.
131 Reader 1993:235.

7 The media and JL123

1 Akhavan-Majid 1990:1007–8.
2 Pressnet 2010.
3 Horsley interview.
4 Freeman 2000:80.
5 Freeman 2000:69–74.
6 Freeman 2000:130–40.
7 Gamble and Watanabe 2004:36.
8 Cobb and Primo 2003:1–2.
9 Asahi Shimbun Shakaibu 2006:72.
10 Horsley interview.
11 Horsley interview.
12 Video Research 1985:3.
13 Asahi Shimbun Shakaibu 2006:98.
14 Video Research 1985:3.
15 Kohira 1991.
16 Interviews with journalists, 2009.
17 Asahi Shimbun Shakaibu 2006:166.
18 Interviews with journalists, 2009.
19 Fujita 2005b:30.
20 *Nikkei Shimbun* 1985-10-15.
21 Asahi Shimbun Shakaibu 2006:105.
22 Earhart 2008:434.
23 Haruta and Hallahan 2003:138.
24 *Mainichi Shimbun* 2010-01-09.
25 Irei-no-Sono Zaidan 1988.
26 Weiner and Chapman 2009:172.
27 Ogawa interview August 2009.
28 Mizushima 2003.
29 Asahi Shimbun Shakaibu 2006:195.
30 Iizuka 2006:238–9.
31 Kobayashi 2010; *izoku* interview.
32 Iizuka 2006:70.
33 Kadota 2010:52.
34 Miyajima interview.
35 Gamble and Watanabe 2004:81.
36 http://sk.fox.ac/jal123/emma1.jpg, http://sk.fox.ac/jal123/focus2.jpg
37 http://sk.fox.ac/jal123/emma2.jpg
38 http://sk.fox.ac/jal123/emma3.jpg, http://sk.fox.ac/jal123/emma7.jpg, http://sk.fox.ac/jal123/focus1.jpg, http://sk.fox.ac/jal123/focus3.jpg, http://sk.fox.ac/jal123/focus4.jpg, http://sk.fox.ac/jal123/friday1.jpg, http://sk.fox.ac/jal123/friday3.jpg, http://sk.fox.ac/jal123/friday4.jpg, http://sk.fox.ac/jal123/jal13.jpg
39 http://sk.fox.ac/jal123/emma4.jpg, http://sk.fox.ac/jal123/emma5.jpg, http://sk.fox.ac/jal123/friday2.jpg
40 http://sk.fox.ac/jal123/emma8.jpg
41 van Wolferen 1989:319–20.
42 Asahi Shimbun Shakaibu 2006:90–1.
43 van Wolferen 1989:319.
44 van Wolferen 1989:465ff.
45 Anonymous interview.
46 Tudor interview.
47 Satō 2005:129.
48 TBS letter 2010-09-09.
49 Walter 1993:69–70.
50 Reader 1993:239.
51 Hood 2008.
52 Ozawa interview.
53 Fukushima interview.
54 Interviews with journalists.

8 Japan's *Titanic*

1 Fairhall 1993:1.
2 Vinitzky-Seroussi 2005:231.
3 Halbwachs 1992.
4 Davie 1986:215.
5 Berlinguez-Kōno 2008:238.
6 Young 1993:9.
7 Vinitzky-Seroussi 2005:235.
8 Maier 1998.
9 Creighton 2009:41.
10 Assmann 2006.
11 Saaler and Schwentker 2008:6.
12 Schäfer 2008:167.
13 Schäfer 2008:166–7.
14 Coates and Morrison 1991:72, 79, 83; van Bremen 2005:38.
15 Ikeda 2004.
16 Ikeda 2004:22-4.
17 Ikeda 1998b.
18 Ikeda 2004:47.
19 Ikeda email 2010-12-01.
20 e.g. Yoshihara 2004:74–80.
21 Anonymous email.
22 Ogawa interview.
23 Ikeda 1998a:38.
24 JAL, untitled press release, 1985-08-13.
25 AAIC 1987b:60.
26 AAIC 1987b:13.
27 *see* BBC 2005a.
28 Yamazaki email.
29 TBS 2005.
30 Bayly interview.
31 *Izoku* interview.
32 Unmeinonazowosagurukai 2008:18.
33 Unmeinonazowosagurukai 2008:19.
34 Unmeinonazowosagurukai 2008:19.

35 Mathews diary and interviews.
36 *Izoku* interviews.
37 e.g., Kawamura 2005:38, 196.
38 Iizuka 2001:62–3.
39 Kawamura 2005:39.
40 Ascari 2009.
41 Yokoyama interview 2009.
42 Yokoyama interview 2009.
43 Walter 1993:73.
44 Miyajima interview 2009.
45 Miyajima interview 2009.
46 Kawamura 2005:208.
47 Honya 2010.
48 Yamazaki 2001:153–6.
49 Wakaizumi interview 2010.
50 Kubo interview 2010.
51 Director's commentary on *Kuraimāzu Hai* (2008) DVD.
52 Miyajima interview 2009.
53 Yokoyama interview 2009; Kubo interview 2010.
54 *Sankei Shimbun* 2009-10-23.
55 *Shinema Hōchi* 2009-11-27.
56 *Shizumanu Taiyō* DVD; Kadokawa email 2008-12-18.
57 *Shizumanu Taiyō* DVD.
58 Lennon and Foley 2004:8.
59 Wakaizumi interview 2010.
60 *Izoku* interviews.
61 TBS 2005.
62 Lennon and Foley 2004:78; Baudrillard 1983.
63 Heinze 2010.
64 Wakaizumi interview 2010.
65 Video Research 2009.
66 Job 1996.
67 Barlay 1990:264–5.
68 Bartlett 2009:127–8.
69 Lennon and Foley 2004:169.
70 BBC 2008c.
71 Books LLC 2010.
72 Antonucci 1995.
73 BBC 1999a.
74 Antonucci 1995.
75 Kawamura 2004:46–50.
76 Kawamura 2004:49–50.
77 Yoneda 2005:104–21.
78 Cineflix 2005.
79 AAIC 1987b:117.
80 Lennon and Foley 2004:6.
81 Data from Google Analytics.
82 Lennon and Foley 2004:3.
83 Lennon and Foley 2004:8.
84 Lennon and Foley 2004:8.
85 Lennon and Foley 2004:11.
86 Lennon and Foley 2004:119–20.
87 Walter 2009:54.
88 Reader 1993:236.
89 Creighton 2009:44.
90 Lennon and Foley 2004:168.
91 *Nikkei Shimbun*, 1986-01-28.
92 Lennon and Foley 2004:61.
93 West Sellars and Walter 1991:184–5, 195, 196.
94 West Sellars and Walter 1993:195.
95 Lennon and Foley 2004:120.
96 Lennon and Foley 2004:58.
97 Hood 2006:146.
98 JAL interview 2007.
99 Lennon and Foley 2004:166.
100 Lennon and Foley 2004:10, 145.
101 Hood 2006:167.
102 Lennon and Foley 2004:18, 165.
103 Schäfer 2008:166–7.
104 Miyajima and Ozawa interview 2009.
105 Nakamaki 2005:45.

9 Conclusions: JL123 present and future

1 *The Japan Times*, 2010-07-14.
2 Miyajima interview.
3 Katō 2005:231.
4 Katō 2005:204–12.
5 Taylor email 2008-04-02.
6 AAIC 1987b:106.
7 AAIC 1987b:106.
8 Yoshioka 1986:279–80.
9 Kawamura 2004:233.
10 Yoshioka 1986:311.
11 e.g. Nihon no Kuroi Kiri 2010.
12 Aoyama 2010:153.
13 Anonymous interview.
14 Hooper McCarty and Foecke 2008:189-90
15 Hooper McCarty and Foecke 2008:95.
16 Weir 2000:239.
17 *Yomiuri Shimbun* 2009-11-19.
18 Takamoto interview.
19 AAIB email 2009-02-13.
20 Takamoto interview.
21 FAA letter, 1985-09-25.
22 Kawakita interview.
23 JAL 2010.
24 ARAIC 2002:2, 134.
25 *Mainichi Shimbun* 2010-03-10.
26 *Mainichi Shimbun* 2010-11-26.

27 Anonymous interviews.
28 Ministry of Defense and Fire and Disaster Management Agency interviews.
29 *Mainichi Shimbun* 2010-01-12.
30 *Mainichi Shimbun* 2010-03-01.
31 *Mainichi Shimbun* evening edition 2008-02-19.
32 *Mainichi Shimbun* 2008-11-29.
33 *Mainichi Shimbun* 2008-11-14.
34 Ministry of Defense 2010.
35 *Asahi Shimbun* 2008-05-30.
36 Miyajima interview.
37 Reader 1993:234.
38 Reader 1993:237.
39 Reader 1993:125, 237; Reader 2005:89, 91.
40 Guichard-Anguis 2009:4.
41 Walter 1993:72.
42 JAL 2002:CD-ROM.
43 *Yomiuri Shimbun* 2009-10-30.
44 *Mainichi Shimbun* 2010-08-03.
45 JAL 2011.
46 *The Japan Times* 2011-03-01.
47 Lennon and Foley 2004:25.
48 Walter 1993:73.
49 *Asahi Shimbun* 1995-08-12.
50 *Yomiuri Shimbun* 2000-08-12.
51 *Asahi Shimbun* 1995-08-12.
52 *Sankei Shimbun* 2005-08-12.
53 *Nikkei Shimbun* 2010-08-06.
54 Kosuge interview; Science Daily 2007.
55 Kawahara interview.
56 Allcock 1993.
57 West Sellars and Walter 1993.
58 Lennon and Foley 2004:163.
59 Nora 1996:7.
60 Schäfer 2008:166.

Bibliography

To save space, and as many sources that I have read often cover similar material, I have tried to limit the number of sources referred to. Similarly, this bibliography only contains those materials which have been cited in the text and a selection of other significant works.

8/12 Renrakukai (1987) *Osutaka Rekuiemu*, Tōkyō: Mainichi Shimbun.

— (1991) *Futatabi no Osutaka Rekuiemu*, Tōkyō: Mainichi Shimbun.

— (2005) *Akanegumo Sōshūhen – Nikkōki Osutaka-yama Tsuiraku Jiko Izoku no 20 Nen*, Tōkyō: Honnoizumisha.

— (2006) *Tabiji – Shijitsu-o Motomete*, Maebashi: Jōmō Shimbunsha.

AAIC (Aircraft Accident Investigation Commission, Ministry of Transport) (1987a) *Aircraft Accident Investigation Report: JA8119* (Japanese version). Available at jtsb.mlit.go.jp/jtsb/aircraft/download/pdf/62-2-JA8119.pdf, accessed 1 February 2007.

— (1987b) *Aircraft Accident Investigation Report: JA8119* (English version). Available at www.mlit.go.jp/jtsb/eng-air_report/JA8119.pdf, accessed 1 February 2007.

ACRO (Aircraft Crashes Record Office) (2010a) 'Death Rate per Year', www.baaa-acro.com/Liste%20des%20deces%20par%20annee.htm, accessed 8 March 2010.

— (2010b) 'Statistics', www.baaa-acro.com/Statistiques%20diverses.htm, accessed 8 March 2010.

— (2010c) 'Japan', www.baaa-acro.com/Pays/J/Japon.htm, accessed 8 March 2010.

— (2010d) 'Worst Accidents List', www.baaa-acro.com/Liste%20des%20accidents.htm, accessed 1 September 2010.

Akhavan-Majid, R. (1990) 'The press as an elite', *Journalism Quarterly*, vol. 67, no. 4 (Winter 1990), pp. 1006–14.

Alexander, D. (2005) 'An interpretation of disaster in terms of changes in culture, society and international relations' in Perry, R.W., Quarantelli, E.L. (eds) *What is a Disaster?* Philadelphia: Xlibris, 25–38.

Allcock, J. (1993) 'Kosovo: The Heavenly and the earthly crown' in Reader, I and Walter, T. (eds) *Pilgrimage in Popular Culture*, Basingstoke: Macmillan.

Altman, K.K. (1996) 'Television and political turmoil' in Pharr, S.J. and Krauss, E.S. (eds) (1996) *Media and Politics in Japan*, Honolulu: University of Hawai'i Press.

Anderson, D.F. and Eberhardt, S. (2010) *Understanding Flight*, McGraw-Hill.

Antonucci, M. (1995) '1985 air crash rescue botched, ex-airman says', *Pacific Stars and Stripes*, 27 August 1995, pp. 1 and 6.

Aoki, E., Imashiro, M., Kato, S. and Wakuda, Y. (2000) *A History of Japanese Railways 1872–1999*, Tōkyō: East Japan Railway Culture Foundation.

Aoki, T. (1994) *Ontake Junrei*, Tōkyō: Kōdansha.

Aoki, Y. (2005) *Jiko Chōsa Hōkokusho-ga Kataru Kōkū Jiko no Shinjitsu*, Tōkyō: Ikarosu.

Aoyama, T. (2010) *Tenkū no Hoshi-tachi-e*, Tōkyō: Magajinrando.

Arai, M. (2004) *Sen-no-Kaze-ni Natte*, Tōkyō: Rironsha.

ARAIC (Aircraft and Railway Accidents Investigation Commission) (2002) Air Accident Investigation Report Japan Airlines Flight 907 and Japan Airlines Flight 958, available at www.mlit.go.jp/jtsb/eng-air_report/JA8904.pdf (English), accessed 1 October 2010.

Asahi Shimbun English Web Edition (2005) 'Weekend beat: "thoroughbread yakuza" survives suspicion, shootout', www.asahi.com/english/Herald-asahi/TKY200510220177.html, accessed 23 October 2005.

Asahi Shimbun Nippon Jinmyakukihan (2008) *Watashi no Karada no Mama Daite*, Tōkyō: Asahi Shimbun.

Asahi Shimbun Shakaibu (2006) *Nikkō Janboki Tsuiraku: Asahi Shimbun no 24ji*, (17th edition, first published 1990) Tōkyō: Asahi Shimbun.

Asahi Shimbun *(2003) Japan Almanac 2004, Tōkyō: Asahi Shimbun.*

— *(2010)* adv.asahi.com/2011/004.pdf, accessed 1 December 2010.

Ascari, M., 'Reading world literature: narratives and Intercultural Understanding', Cardiff University workshop, 26 October 2009.

ASN (Aviation Safety Network) (2010a) 'ASN Aircraft Accident Boeing 727-89 JA8315', aviation-safety.net/database/record.php?id=19700331-0, accessed 8 March 2010.

— (2010b) 'ASN Aircraft Accident Boeing 727 Tokyo', aviation-safety.net/database/record.php?id=19721106-0, accessed 8 March 2010.

— (2010c) 'ASN Aircraft Accident Boeing 747-246B JA8109', aviation-safety.net/database/record.php?id=19730723-0, accessed 8 March 2010.

— (2010d) 'ASN Aircraft Accident Boeing 747 Naha', aviation-safety.net/database/record.php?id=19740312-0, accessed 8 March 2010.

— (2010e) 'ASN Aircraft Accident Boeing 747 Tokyo', aviation-safety.net/database/record.php?id=19750409-1, accessed 8 March 2010.

— (2010f) 'ASN Aircraft Accident McDonnell Douglas DC-8 Manila', aviation-safety.net/database/record.php?id=19760105-0, accessed 8 March 2010.

— (2010g) 'ASN Aircraft Accident McDonnell Douglas DC-10-40', aviation-safety.net/database/record.php?id=19791113-2, accessed 8 March 2010.

— (2010h) 'ASN Aircraft Accident Douglas DC-8-53 JA8012', aviation-safety.net/database/record.php?id=19720614-1, accessed 8 March 2010.

— (2010i) 'ASN Aircraft Accident Douglas DC-8-53 JA8013', aviation-safety.net/database/record.php?id=19720924-1, accessed 8 March 2010.

— (2010j) 'ASN Aircraft Accident McDonnell Douglas DC-8-62 JA8040', aviation-safety.net/database/record.php?id=19721128-1, accessed 8 March 2010.

— (2010k) 'ASN Aircraft Accident McDonnell Douglas DC-8-62H JA8051', aviation-safety.net/database/record.php?id=19770927-0, accessed 8 March 2010.

— (2010l) 'ASN Aircraft Accident McDonnell Douglas DC-8', aviation-safety.net/database/record.php?id=19770928-0, accessed 8 March 2010.

— (2010m) 'ASN Aircraft Accident McDonnell Douglas DC-8-61 JA8061', aviation-safety.net/database/record.php?id=19820209-0, accessed 8 March 2010.

Assmann, J. (2006) *Religion and Cultural Memory*, Stanford: Stanford University Press.

Barlay, S. (1990) *The Final Call: Why Airline Disasters Continue to Happen*, New York: Pantheon Books.

Bartlett, C. (2009) *Air Crashes and Miracle Landings*, Buckinghamshire: OpenHatch Books.

Baudrillard, J. (1983) *Simulations*, New York: Semiotext.

*BBC (*1999a) *Disaster: JAL 123 – A Japanese Tragedy.*

— (1999b) 'Orbiter loss blamed on "silly mistakes"', news.bbc.co.uk/1/hi/sci/tech/514763.stm, accessed 12 November 1999.

— (2001) 'Japan's land prices slide on', news.bbc.co.uk/1/hi/business/1472258.stm, accessed 10 May 2010.

— (2003) '1985: The black ball final' news.bbc.co.uk/sport1/hi/other_sports/snooker/2953933.stm, accessed 1 May 2010.

— (2004a) 'Nine die in Japan "suicide pacts"', news.bbc.co.uk/1/hi/world/asia-pacific/3735372.stm, accessed 1 May 2009.

— (2004b) 'Japan's internet "suicide clubs"', news.bbc.co.uk/1/hi/programmes/newsnight/4071805.stm, accessed 1 May 2009.

— (2005a) '1985: Hundreds dead in Boeing crash', news.bbc.co.uk/onthisday/hi/dates/stories/august/12/newsid_2529000/2529031.stm, accessed 1 February 2007.

— (2005b) *Hiroshima.*

— (2005c) 'Mobile rack up 20 years of use', news.bbc.co.uk/1/hi/technology/4138449.stm, accessed 16 June 2010.

— (2006) 'Timeline: Nuclear plant accidents', news.bbc.co.uk/1/hi/sci/tech/5165736.stm, accessed 11 July 2006.

— (2008a) 'Hole forces Qantas plane to land', news.bbc.co.uk/1/hi/world/asia-pacific/7524733.stm, accessed 25 July 2008.

— (2008b) 'Forced landing for Ryanair flight', news.bbc.co.uk/1/hi/uk/7581492.stm, accessed 26 August 2008.

— (2008c) 'Warning sounded on web's future', news.bbc.co.uk/1/hi/technology/7613201.stm, accessed 15 September 2008.

— (2009) 'Heathrow crash passengers to sue', news.bbc.co.uk/1/hi/england/london/8368521.stm., accessed 19 November 2009.

— (2010) 'Are children more likely to survive plane crashes', news.bbc.co.uk/1/hi/magazine/8679572.stm, accessed 13 May 2010.

Befu, H. (1983) 'Internationalization of Japan and *Nihon Bunkaron*' in Mannari, H. and Befu, H. (1983) *The Challenge of Japan's Internationalization: Organization and Culture*, Kōbe: Kwansei Gakuin University, 232–66.

Benedict, R. (1946) *The Chrysanthemum and the Sword*, Cleveland: Meridian Books.

Berlinguez-Kōno, N. (2008) 'How did Saigō Takamori become a national hero after his death? The political uses of Saigō's figure and the interpretation of *Seikanron'* in Saaler, S. and Schwentker, W. (eds) (2008) *The Power of Memory in Modern Japan, Folkestone: Global Oriental.*

Bernstein, A. (2006) *Modern Passings: Death Rites, Politics, and Social Change in Imperial Japan*, Hawai'i: University of Hawai'i Press.

Boeing (2009) *Statistical Summary of Commercial Jet Airplane Accidents*, www.boeing.com/news/techissues/pdf/statsum.pdf, accessed 1 October 2009.

Books LLC (2010) *Aviation Accidents and Incidents in Japan*, Memphis: Books LLC.

Bowman, M. (1993) 'Drawn to Glastonbury' in Reader, I and Walter, T. (eds) (1993) *Pilgrimage in Popular Culture,* Basingstoke: Macmillan.

Browne, R.B. and Browne, P. (eds) (1991) *Digging into Popular Culture*, Ohio: Bowling Green State University Popular Press.

Brus, M. (1999) 'In the event of a water landing', www.slate.com/id/1003275, accessed 8 October 2010.

Cabinet Secretariat (2010) 'The Government Aircraft', www.kantei.go.jp/foreign/vt2/main/07/photo-senyoki01.html, accessed 3 March 2010.

— (2011) '*Heisei 23-Nen (2011) Tōhoku Chihō Taiheiyōoki Jishin (Higashi Nihon Daishinsai) Nitsuite*', www.kantei.go.jp/saigai/pdf/201104171700jisin.pdf, accessed 18 April 2011.

Channel 4, *Black Box: A Wing and A Prayer.*

Chūgoku Shimbunsha (ed.) (1986) *Nenpyō Hiroshima 40 nen no kiroku*, Hiroshima: Chūgoku Shimbunsha.

Chiba, H. (1997) 'So you want to be a volunteer? The rise of corporate and governmental support for voluntarism in Japan', *Look Japan*, 43, 396, 4–9.

Cineflix (2005) *Out of Control.*

Coates, K.S. and Morrison, W.R. (1991) 'Towards a methodology of disasters: the case of the *Princess Sophia*', in Browne, R.B and Browne, P. (eds) (1991) *Digging into Popular Culture: Theories and Methodologies in Archaeology, Anthropology, and Other Fields*, Ohio: Bowling Green State University Press, 71–83.

Cobb, R.W. and Primo, D.M. (2003) *The Plane Truth: Airline Crashes, the Media and Transportation Policy*, Washington: Brookings Institution.

Creighton, M. (2009) 'The heroic Edo-ic' in Guichard-Anguis, S. and Moon, O. (eds) (2009) *Japanese Tourism and Travel Culture*, London: Routledge.

Cutter, S.L. (2005) 'Are we asking the right question?' in Perry, R.W., Quarantelli, E.L. (eds) (2005) *What is a Disaster?*, Philadelphia: Xlibris, 39–48.

Davie, G. (1993) 'You'll never walk alone: the Anfield pilgrimage' in Reader, I. and Walter, T. (eds) (1993) *Pilgrimage in Popular Culture*, Basingstoke: Macmillan.

Davie, M. (1986) *The Titanic: The Full Story of a Tragedy*, London: Bodley Head.

Dumbrell, J. (2001) *A Special Relationship*, London: Macmillan Press.

Earhart, D.C. (2008) *Certain Victory: Images of World War II in the Japanese Media*, New York: M.E. Sharpe.

Earhart, H.B. (1974) *Religion and the Japanese Experience*, Encino: Dickenson Publishing.

Endow, A. (2003) *France, Germany and the European Union*, Delhi: Aakar Books.

FDMA (Fire and Disaster Management Agency) (2010) www.e-college.fdma.go.jp/husuigai/01/index01.html, accessed 24 March 2010.

Fairhall, J. (1993) *James Joyce and the question of history*, Cambridge: University of Cambridge press.

Faith, N. (1997) *Black Box*, London: Boxtree.

Ferguson, J. (1981) *A Book of Disasters*, London: Hippo.

— (1988) *Modern Disasters*, London: Hippo.

Freeman, L.A. (2000) *Closing the shop: Information caretels and Japan's mass media*, Princeton: Princeton University Press.

Fujita, H. (2005a) *Ano Kōkūki Jiko-wa Kōshite Okita*, Tōkyō: Shinchōsha.

— (2005b) *Kakusareta Shōgen: Nikkō 123bin Tsuiraku Jiko*, Tōkyō: Shinchosha.

Fuji TV (2005) *8/12 Nikkō Tuiraku Jiko 20nen me no Chikai.*

Gamble, A. and Watanabe, T. (2004) *A Public Betrayed: Japanese Media Atrocities, What the World Needs to Know*, Washington: Regnery Publishing.

Gero, D. (2006) *Aviation Disasters: The World's Major Civil Airliner Crashes Since 1940*, 4th Edition, Sparkford: Haynes.

Goo (2010a) *Tōkyō-no Kako-no Tenki*, weather.goo.ne.jp/past/1985/08/662/index.html, accessed 5 October 2010.

— (2010b) *Maebashi-no Kako-no Tenki,* weather.goo.ne.jp/past/1985/08/624/index.html, accessed 22 October 2010.

Gotō, M. (2005) *Watashi-dake no Kunshō*, Tōkyō: Iwanami Shoten.

Gotō, T. (2008) *Nihon Kōkū Kusatta Taiyō: Tatta Hitori no Kakumeiji*, Tōkyō: Otowa.

Guichard-Anguis, S. (2009) 'Introduction' in Guichard-Anguis, S. and Moon, O. (eds) (2009) *Japanese Tourism and Travel Culture*, London: Routledge.

Gunma-ken Ishikai (1986) *Nikkō 123bin Jiko to Ishikai no Katsudō*, Maebashi: Gunma-ken Ishikai.

Gunma-ken Sōmubu (1986) *Nikkō 123bin Tsuirakujiko Taisaku no Kiroku.*

Hadfield, P. (2001) 'Japan struggles with soaring death toll in Suicide Forest', *Daily Telegraph*, 16 June.

Halbwachs, M. (1992) *On Collective Memory*, Chicago: University of Chicago Press.

Hammer, J. (2006) *Yokohama Burning: The Deadly 1923 Earthquake and Fire that Helped Forge the Path to World War II*, New York: Free Press.

Hanazawa, Y. (2008) *(CAP)...*, Tōkyō: Gihōdō.

Haruta, A. and Hallahan, K. (2003) 'Cultural issues in airline crisis communications: A Japan-US comparative study', Asian Journal of Communication, 13:2, 122–50.

Heinze, U (2010) 'Nationalism as Nostalgia in Ishihara Shintarō's Kamikaze film Ore (2007)', Cardiff Japanese Studies Seminar, Cardiff University, 11 February.

Hirooka, Y. (2006) *JAL-ga Abunai*, Tōkyō: Eeru.

Hoffmann, S. and Oliver-Smith, A. (eds) (2002) *Catastrophe and Culture: The Anthropology of Disaster*, Sante Fe: School of American Research Press.

Honya (2010) Honya Taishō, www.hontai.or.jp/history/index.html, accessed 5 September 2010.

Hood, C.P. (2001) *Japanese Education Reform: Nakasone's Legacy*, London: Routledge.

— (2002) 'Images of Japan/Japan 2001 Survey Report', available at www.hood-online.co.uk/publications/REPORT.PDF.

— (2006) *Shinkansen: From Bullet Train to Symbol of Modern Japan*, London: Routledge.

— (2008) 'Researching Japan's Worst Plane Crash', *JAWS (Japan Anthropology Workshop)* Newsletter, no. 42, April 2008, ISSN 1356-5631, 38–40.

— (2010) 'Flying experiences', www.hood-online.co.uk/publications/flying.php

— (2012) 'Disaster and Death in Japan: Responses to the Flight JL123 Crash' in Suzuki, H. (ed.) (2012) *Death, Dying and Disposal in Contemporary Japan*, London: Routledge.

Hooper McCarty, J. and Foecke, T. (2008) *What Really Sank the Titanic: New Forensic Discoveries*, New York: Citadel Press.

Horsley, W. and Buckley, R. (1990) *Nippon: New Superpower*, London: BBC Books.

Hui, T.Y. (2005) 'The Violent and the Benign: How Kōbe Remembers its Rivers' in Hui, T.Y., van Bremen, J. and Ben-Ari, E. (eds) (2005) *Perspectives on Social Memory in Japan*, Folkestone: Global Oriental.

Hui, T.Y., van Bremen, J. and Ben-Ari, E. (eds) (2005) *Perspectives on Social Memory in Japan*, Folkestone: Global Oriental.

Iizuka, S. (2001) *Tsuiraku Itai – Osutaka-yama no Nikkōki 123bin*, Tōkyō: Kōdansha.

— (2005) *Tsuiraku Gemba Nokosareta Hitotachi – Osutaka-yama, Nikkōki 123bin no Shinjitsu*, Tōkyō: Kōdansha.

Ikeda, C. (1996) *Yukidoke no One – Nikkōki Jiko kara 11nen*, Tōkyō: Bungeisha.

Ikeda, M. (1998a) *Osutaka-yama Fairu JAL123bin Tsuiraku 'Jiko' Shinsō Kaimei*, Tōkyō: Bungeisha.

— (1998b) *JAL123bin-wa Jieitai-ga Gekitsui Shita,* Tōkyō: Bungeisha.

— (1999) *Osutaka-yama Fairu 3 JAL123bin Kūhaku no 14 Jikan*, Tōkyō: Bungeisha.

— (2001) *Intānetto-de Toku JAL123bin Jiken*, Tōkyō: Bungeisha.

— (2004) *Kanzen Hanzai: JAL123bin Gekitsui Jiken*, Tōkyō: Bungeisha.

Ino, Y. (2009) *Manga to Jūdai Shōgen de Kanzen Suiri: Shōwa/Heisei Kōrudokēsu*, Tōkyō: Takarajimasha.

Inoue, T. (1985) *Kōkū Daijiko no Yosoku: Nikkō Jamboki Tsuiraku no Nazo*, Tōkyō: Tairikushobō.

International Federation of Air Line Pilots Associations (IFALPA), no date, 'Regarding Crash Investigation of JL123', copy of statement received from JAL.

IREI Air Safety (1999) *Kōkū Anzen Kokusai Rarī*.

Irei-no-Sono Zaidan (1988) *Chinkon-no Shiori*, Gunma: Irei-no-Sono Zaidan.

— (1991) *Osutaka no Sōkatachi: The Flowers of Osutaka, Gunma:* Irei-no-Sono Zaidan.

Ishibuchi, Y. (ed.) (2003) *Uenomurashi (VIII) Uenomura no Rekishi*, Gunma: Ueno-mura.

Ishida, J. and Takeda, R. (1985) *Meisō – Dokyumento Nihon Kōkū*, Tōkyō: Keirinshobō.

Ishihara, S. (1989) *The Japan That Can Say No*, trans. F. Baldwin, New York: Simon & Schuster.

Iwaksaka, M. and Toelken, B. (1994) *Ghosts and the Japanese: Cultural Experience in Japanese Death Legends*, Utah: Utah State University Press.

JAL (1985a) 'A View of JAL in the New Aviation Era', JAL memo.

— (1985b) '747SR Background Data', memo from JAL Tōkyō to JAL London.

— (1985c) 'Suicide Case', JAL memo.

— (1985d) 'What the Foreign Media Said', JAL memo.

— (1985e) 'Omimai', JAL memo.

— (1985f) 'Statement made by Mr Takayuki Hashizume, Senior Vice President of Japan Air Lines, regarding compensation in the case of the accident of August 12', JAL statement.

— (1985g) 'Time for Courage', *Ohzora*, Tōkyō: Japan Airlines.

— (1988) 'Japan Air Lines Return to the Private Sector', JAL memo.

— (2001) 'JAL and Boeing – Ms. Susanne Bayly; A statement from Japan Airlines', 10 August 2001.

— (2002) 'Contrail of JAL Group 1951-2001', Book and CD-ROM, Tōkyō: JAL.

— (2004) 'World Top 11 City-Pairs', SORA Data File.

— (2010) 'Safety Operations', www.jal.com/en/safety/info/info4.html, accessed 17 November 2010.

— (2011) 'JAL Group Re-establishes Identity with Corporate Policy and Logo Change', press.jal.co.jp/en/release/201101/001728.html, accessed 19 January 2011.

JMA (Japan Meteorological Agency) (2010a) '*Kako no Jishin/Tsunami Higai*', www.seisvol.kishou.go.jp/eq/higai/higai-1995.html, accessed 11 March 2010.

— (2010b) www.jma.go.jp/jma/kishou/know/typhoon/6-1.html, accessed 25 March 2010.

— (2010c) www.jma.go.jp/jma/press/0412/20a/tenko2004.pdf, accessed 25 March 2010.

— (2010d) www.data.kishou.go.jp/climate/cpdinfo/climate_change/2005/pdf/2005_1-1.pdf, accessed 25 March 2010.

— (2010e) '*Shindo ni tsuite*', www.jma.go.jp/jma/kishou/know/shindo/shindokai.html, accessed 26 March 2010.

— (2010f) '*Daini Miyakojima Taifū*', www.data.jma.go.jp/obd/stats/data/bosai/report/1966/19660904/19660904.html, accessed 26 March 2010.

— (2010g) '*Nihon Tsukinde Hassei Shita Omona Higaijishin*', www.seisvol.kishou.go.jp/eq/higai/higai1996-new.html, accessed 1 September 2010.

Job, M. (1995) *Air Disaster: Volume 1*, Fyshwick: Aerospace Publications.

— (1996) *Air Disaster: Volume 2*, Fyshwick: Aerospace Publications.

Job, M. and Swift, S. (2005) 'JAL123: August 1985', *Flight Safety Australia*, July–August 2005, 28–33.

Kadota, R. (2010) *Kaze-ni soyogu bohyō*, Tōkyō: Shūeisha.

Kakehashi, K. (2009) *Shōwa-no Isho*, Tōkyō: Bungei Shunjū.
Kashiwagi, Y. (1986) *Ue-o Muite Arukō*, Tōkyō: Fuji TV Shuppan.
Katō, K. (2001) *Tsuiraku: Kyōgaku no Shinjitsu*, vol. 1, Tōkyō: Kōdansha.
— (2005) *Kowareta Biyoku – Nikkō Jamboki Tsuiraku no Shinjitsu*, Tōkyō: Kōdansha.
— (2006a) *Renzoku Tsuiraku Saishin'eiki – Kōkūki Jiko, Fukuzatsu Kaiki Nari*, Tōkyō: Daiwa.
— (2006b) *Bakuhatsu JAL123bin – Kōkūki Jiko, Fukuzatsu Kaiki Nari*, Tōkyō: Daiwa.
— (2007) *Masaka no Tsuiraku*, Tōkyō: Daiwa.
— (2008) *Kōkūki Jiko 50nenshi*, Tōkyō: Kōdansha.
Katō, Y. (2009) *Watanabe Ken Shuen Eiga Shizumanu Taiyō Dokyumentobukku*, Tōkyō: Kadokawa Gurūpu Paburishingu.
Kawakita, T. (1992) *Tsuiraku no Ato*, Tōkyō: Bungei Shunjū.
Kawamura, K. (2004) *Nikkōki Tsuiraku*, Tōkyō: Īsuto Puresu.
— (2005) *Nikkōki Itai Shūyō – 123bin, Jiko Shori no Shinsō*, Tōkyō: Īsuto Puresu.
King, C. (1993) 'His Truth Goes Marching On: Elvis Presley and the Pilgrimage to Graceland' in Reader, I. and Walter, T. (eds) (1993) *Pilgrimage in Popular Culture*, Basingstoke: Macmillan.
Kingston, J. (2004) *Japan's Quiet Transformation*, London: Routledge.
Kitagawa, J.M. (1987) *On Understanding Japanese Religion*, Princeton: Princeton University Press.
Kiyono, R. (1990) *Osutaka-yama*, Tōkyō: Kindaibungeisha.
Kluckhohn, C. (1949) *Mirror for Man*, New York: McGraw-Hill.
Kobayashi, M. (2010) *Osutaka-yama-no Atsui Natsu*, Tōkyō: Genbun Magazine.
Kōdansha (2005) *Shōwa Nippon: vol. 23: Kagaku Banpaku to Nikkō Osutaka-yama Tsuiraku – 120,000,000 nin no Eizō*, DVD and Book, Tōkyō: Kōdansha.
— (2010) Kōdansha Nonfikushon-shō, www.kodansha.co.jp/award/archive/nonfiction.html, accessed 17 November 2010.
Kohira, N. (1991) *4/524*, Tōkyō: Shinchōsha.
Kokubu, H.S. (2008) 'Japan Airlines Flight 123', qhdc.org/jal123.htm, accessed on 1 June 2010.
Komatsu, K. (2000) *Kioku suru minzokugaku*, Tōkyō: Jimbun Shoin.
— (2002) *Kaminaki jidai no minzokugaku*, Tōkyō: Serika Shobō.
Kubo, M. (2009) *JAL Jet Story*, Tōkyō: Ikarosu Shuppan.
Lennon, J. and Foley, M. (2004) *Dark Tourism*, London: Thomson Learning.
Levinson, J. and Granot, H. (2002) *Transportation Disaster Response Handbook*, London: Academic Press.
Maier, C.S. (1998) *The Unmasterable Past: History, Holocaust and German National Identity*, Cambridge, MA: Harvard University Press.
Mainichi Shimbun (2010) macs.mainichi.co.jp/now/section-b/01.html, accessed 1 December 2010.
Mathews, P. and Haines, K. (2007) *JL123: Personal Experiences of the Aftermath of the World's Worst Plane Crash*, Cardiff Japanese Studies Seminar, 20 March 2007.
Ministry of Defense (2010) 'International Disaster Relief Operations', www.mod.go.jp/e/d_act/ipca/idro.html, accessed 17 November 2010.
Miyajima, K. (1988) *Itsumademo Isshodayo – Nikkō Jambo Osutaka-yama Tsuiraku Jiko de itta Ken-chan e*, Tōkyō: Fusōsha.
— (1989) *Shiroi Koinobori*, Tōkyō: Nobinokai.
— (2010) *Osutaka-yama-to ikiru*, Tōkyō: Shinchōsha.
Miyata, N. (1988) *Miroku shinkō no kenkyū*, Tōkyō: Miraisha.

Miyake, Y. (1999) The Japanese View of Corpses: The Sacredness of Things, www.relnet.co.jp/relnet/brief/r12-50.htm, accessed 24 April 2007.

Miyamura, H. (2003) *Hōmuri Sarareta Shinjitsu – Nikkō Jamboki Tsuiraku Jiko no Giwaku*, Tōkyō: Seishinsha.

Miyazaki, M. (2005) *Toppamono*, Tōkyō: Kotan Publishing.

Mizushima, A. (2003) 'Puraibashii-ken/shōzo-ken', web.archive.org/web/20070820191114/mizushima-s.pos.to/lecture/2003/030507/030507_06.htm, accessed 6 July 2010.

MLIT (2010a) 'History of Japanese Roads', www.mlit.go.jp/road/road_e/hist/hist1.html, accessed 17 June 2010.

— (2010b) *'Zenkoku Kūkō Haichizu'*, www.mlit.go.jp/koku/04_outline/01_haichi/index.html, accessed 14 September 2010.

— (2010c) *Kyoritaibetsu no Kōtsūkikan Yusōbuntanritsu no Suī*, www.mlit.go.jp/toukeijouhou/unyugeturei/unyugeturei00/topics00/topics12-2-1_.htm, accessed 15 September 2010.

— (2010d) *Kyoritaibetsu no Kōtsūkikan Yusōbuntanritsu no Suī*, www.mlit.go.jp/hakusyo/mlit/hakutsho/h20/excel/j1201220.xls, accessed 15 September 2010.

Moore, A. (1980) 'Walt Disney World: Bounded Ritual Space and the Playful Pilgrimage Center', *Anthropological Quarterly*, vol. 53, no. 4, 207–17.

Morishita, R. (1991) *Akatsuki no Sato – Osutaka-no-one ni Nemuru Haha-e Sasageru*, Tōkyō: Bun'ensha.

Muraoka, N. (2002) *Tenmei no Chinjō – Nikkō Jamboki Osutaka-yama Tsuiraku to Guriko Morinaga Jiken tono Kanren*, Tōkyō: Tōkyōtosho.

Naitō, H. (2005) 'Night Trains I Have Ridden', Bullet-in, Japanese Railway Society, Issue 54, April–June 2005, 15–20.

Naitō, I. (1994) *Shinsetsu: Nihon Kōkūki Jikobo*, Tōkyō: Akishobō.

Nakamaki, H. (2005) 'Memorial Monuments of Interrupted Lives in Modern Japan: From *Ex Post Facto* Treatment to Intensification Devices' in Hui, T.Y., van Bremen, J. and Ben-Ari, E. (eds) (2005) *Perspectives on Social Memory in Japan*, Folkestone: Global Oriental.

Nakano, L. (2005) *Community Volunteers in Japan: Everyday stories of social change*, London: RoutledgeCurzon.

National Defense Counsel for Victims of Karōshi (1990) *Karōshi*, Tōkyō: Mado-sha.

New York Times (1988) 'Japanese Ask Which Side Navy is On', www.nytimes.com/1988/07/30/world/japanese-ask-which-side-navy-is-on.html, accessed 20 October 2010.

— (2008) 'Hole in fuselage forces Qantas 747 to land', www.nytimes.com/2008/07/26/world/asia/26qantas.html?_r=1&hp, accessed 26 July 2008.

Newsweek (1985) 'What went wrong?', 26 August 1985.

NHK (2001) *Nihon Kōkūshi 4*, DVD.

Nichijōren (1994) *Nikkō123bin-ni Kyūgen'atsu-wa Nakatta*, also available at http://www.alpajapan.org/kannkoubutu/genatsu/MOKUJI~1.HTM, accessed 1 April 2007.

Nihon no Kuroi Kiri (2010) 'JAL123 tsuiraku jiko – shinsō-o ō', blog.goo.ne.jp/adoi/e/2ae393ac6bf8cc30807ba140c33183f8, accessed 1 March 2010.

Nikkei Shimbun (2010) adweb.nikkei.co.jp/paper/data/mo/chapter1/01.html, accessed 1 December 2010.

Noda, M. (1992) *Mo-no Tojō nite – Daijiko Izoku no Hiai no Kenkyū*, Tōkyō: Iwanami Shoten.

Noguchi, T. and Fujii, T. (2000) 'Minimizing the Effect of Natural Disasters', *Japan Railway and Transport Review*, 23, 52–9.

Nora, P. (1996) *Realms of Memory*, New York: Columbia University Press.

Notoji, M. (1990) *Dizunirando to iu seichi*, Tōkyō: Iwanami Shinsho.

NPA (National Police Agency) (2003) 'Movements of the Japanese Red Army and the "Yodo-go" Group, www.npa.go.jp/keibi/kokutero1/english/pdf/sec03.pdf, accessed 9 March 2010.

Oedewald, M. (2009) 'Meanings of tradition in contemporary Japanese domestic tourism' in Guichard-Anguis, S. and Moon, O. (eds) (2009) *Japanese Tourism and Travel Culture*, London: Routledge.

Ōkuni, T. (2001) *Mimoto Kakuni: Ha ya Hone kara Apurōchi*, Tōkyō: Furīpuresu.

Ōshima, S. (2009) *Shindara oshimai*, Tōkyō: PHP Kenkyūjo.

Ōuchi, K. (2003) *Nihon no Kōkūki Jiko 90nen*, Tōkyō: Kōtsūkenkyūkai.

Ōtsuka, E. and Yamazaki, H. (2009) *The Kurosagi Corpse Delivery Service Volume 8*, translation produced by Dark Horse Comics, Milwaukie: Dark Horse Manga.

Owen, J. (1995) 'USFJ has no records on crash', *Pacific Stars and Stripes*, 27 August 1995, p. 7.

Ozaki, K. (1992) *Gendai Jinsai Kokusho – Kyodai Jiko no Sōsa to Jānarizumu*, Tōkyō: San'ichishobō.

Özerdem, A. and Jacoby, T. (2006) *Disaster Management and Civil Society: Earthquake Relief in Japan, Turkey and India*, New York: I.B.Tauris & Co.

Perry, R.W., Quarantelli, E.L. (eds) (2005) *What is a Disaster?*, Philadelphia: Xlibris.

Pharr, S. (1986) 'Japan in 1985: The Nakasone Era Peaks', *Asian Survey*, vol. 26, no. 1, pp. 54–65.

Pharr, S.J. and Krauss, E.S. (eds) (1996) *Media and Politics in Japan*, Honolulu: University of Hawai'i Press.

Pressnet (2010) 'Circulation', www.pressnet.or.jp/english/data_e/01circulation.htm, accessed 1 December 2010.

Reader, I. and Walter, T. (eds) (1993) *Pilgrimage in Popular Culture*, Basingstoke: Macmillan.

Reader, I. and Tanabe, G.J. Jr. (1998) *Practically Religious: Worldly Benefits and the Common Religion of Japan*, Honolulu, University of Hawai'i Press.

Reader, I. (1993) 'Dead to the World: Pilgrims in Shikoku' in Reader, I. and Walter, T. (eds) (1993) *Pilgrimage in Popular Culture*, Basingstoke: Macmillan.

— (2005) *Making Pilgrimages: Meaning and Practice in Shikoku*, Honolulu: University of Hawai'i Press.

Reader, I. andreasen, E. and Stefánsson, F. (1993) *Japanese Religions: Past and Present*, Folkestone: Japan Library.

Rodríguez, H., Quarantelli, E.L. and Dynes, R. (eds) (2007) *Handbook of Disaster Research*, New York: Springer.

Rodríguez del Alisal, M., Ackermann, P. and Martinez, D.L. (eds) (2007) *Pilgrimages and Spiritual Quests in Japan*, London: Routledge.

Saaler, S. and Schwentker, W. (eds) (2008) *The Power of Memory in Modern Japan*, Folkestone: Global Oriental.

Saigai Jōhō Sentā and Nichigai Asoshiētsu Henshūbu (eds) (2007) *Tetsudō, Kōkūki Jiko Zenshi*, Tōkyō: Nichigai Asoshiētsu.

Sankei Shimbun (2010) www.sankei-ad-info.com/chara/t-busuu_1.php, accessed 1 December.

Satō, T. (2005) *Hachigatsu jūgonichi-no shinwa: shūsen kinenbi-no mediagaku,* Tōkyō: Chikuma shinsho.

Schäfer, S. (2008) 'The Hiroshima Peace Memorial Museum and its Exhibition' in Saaler, S. and Schwentker, W. (eds) (2008) *The Power of Memory in Modern Japan*, Folkestone: Global Oriental.

Schreiber, M. (1996) Shocking Crimes of Postwar Japan, Tōkyō: Yenbooks.

Schulz, F.R. (2009) ''Envisaging Death in Socialism: Graves, Gravestones, and their absences in the GDR', presentation at Envisaging Death Workshop, Birmingham University, 26 June.

Science Daily (2007) 'New automated system IDs victims of mass disasters in minutes', www.sciencedaily.com/releases/2007/11/071127105531.htm, accessed 7 November 2007.

Scott, P.D. (2008) 'Civil Society, NGOs, and Spatialized Politics: Mobilizing Public Opinion and the War in Iraq' in Eldridge, R. and Midford, P. (eds) (2008) *Japanese Public Opinion and the War on Terrorism*, Hampshire: Palgrave Macmillan, 185–205.

Seeman, R. (1987) 'The Japan Newsletter' available from www.japanlaw.info/lawletter/jan87/fax.htm, accessed 2 February 2010.

Sharpley, R. and Stone, P.R. (eds) (2009) *The Darker Side of Travel*, Bristol: Channel View.

Shinjinbutsuōraisha (ed.) (2010) *Yukashii Shōwa no Nyūsu Techō*, Tōkyō: Shinjinbutsuōraisha.

Sōmuchō (1987) Shōwa 62-nenpan Tsūshin Hakusho, available at www.soumu.go.jp/johotsusintokei/whitepaper/ja/s62/html/s62a02020100.html, accessed 15 June 2010.

Sōmushō (2010a) *Mobairu tsūshin no sekai*, www.soumu.go.jp/joho_tsusin/kids/mobile/kiki/Handyphone.html, accessed 16 June 2010.

— (2010b) *Tōkei Nenkan 2010*, Tōkyō: Sōmuchō.

Spence, C. (ed.) (2007) *Aeronautical Information Manual/Federal Aviation Regulations 2007*, New York: McGraw-Hill.

Stewart, S. (1988) *Air Disasters*, London: Arrow Books.

Sugimoto, S and Nagai, D. (2009) 'Casualities by the 3 June 1991 pyroclastic flow at Unzen volcano', available at hdl.handle.net/2324/13523, accessed 30 September 2010.

Sugiura, K. (2001) *Gekitotsu! Tōkaidōsen: 'Nozomi' tai kōkū shatoru,* Tōkyō: Sōshisha.

— (2007) *Hoka-ni Ochita Nihon Kōkū – Hatashite Jishu Saiken Dekiru-noka*, Tōkyō: Sōshisha.

Suzuki, H. (2001) *The Price of Death: The Funeral Industry in Contemporary Japan*, Stanford: Stanford University Press.

Suzuki, I. (2012) 'The dead to be prayed: considering the idea of *kami* through the photographs of the dead' in Suzuki, H. (ed.) (2012) *Death, Dying and Disposal in Contemporary Japan*, London: Routledge.

Suzuki, K. (1992) *Shitai-ni Ha ari: Hōshigaku no Gemba*, Tōkyō: Tokumashoten.

Suzuki, K. (1999) *Itai Kantei: Ha-ga Katari Kaketekuru*, Tōkyō: Kōdansha.

Takagi, S. (1989) *Kyodai Jiko no Jidai*, Tōkyō: Kōbundō.

Takenaga, K. (1987) *Yurusarezaru Ririku.*

Tamaru, N. and Reid, D. (1996) *Religion in Japanese Culture*, Tōkyō: Kodansha International.

Tanaka, S. (1988) *Kansei-ni Uttaeru Dōwa Kyōiku: Nikkō Jiko kara Musume no Nokoshite-kureta Mono*, Tōkyō: Akashishoten.

— (1996) *Zoku – Kansei-ni Uttaeru Dōwa Kyōiku*, Tōkyō: Akashi Shoten.

Tanakamaru, K. (2002) *Samayoeru eirei tachi: kuni no mitama ie no hotoke*, Tōkyō: Kashiwashobō.

TBS (2005) *Boisurekōdā: Nokosareta Koe no Kiroku – Jambo-ki Tsuiraku 20-nenme no Shinjitsu.* Television documentary first shown on 12 August 2005.

Thang, L.L. (2005) 'Preserving the Memories of Terror: Kōbe Earthquake Survivors as "Memory Volunteers"', in Hui, T.Y., van Bremen, J. and Ben-Ari, E. (eds) (2005) *Perspectives on Social Memory in Japan*, Folkestone: Global Oriental.

The Economist (2002) 'Help! There's nobody in the cockpit', 21 December 2002, vol. 365, issue 8304, 83–5.
Time (1985) 'The Last Minutes of Flight 123', 26 August.
Tsunoda, S. (1993) *Giwaku JAL123bin Tsuiraku Jiko – Konomamadewa 520 Hashira-wa Tsuburenai*, Tōkyō: Wasedashuppan.
Tsuruoka, K. and Kitamura, Y. (1991) *Higeki no Shinsō: Nikkō Jamboki Jiko Chōsa no 677hi*, Tōkyō: Yomiuri Shimbun.
Turner, V. and Turner, E. (1978) *Image and Pilgrimage in Christian Culture*, Oxford: Blackwell.
Uenomura no Chishi Shippitsusha (ed.) (2003) *Uenomurashi (VII) Uenomura no Chishi*, Gunma: Ueno-mura.
Unmeinonazowosagurukai (ed.) (2008) *Fushigisugiru Kimyōna Hanashi*, Tōkyō: Kawade Shobōshinsha.
van Bremen, J. (2005) 'Monuments for the Untimely Dead or the Objectification of Social Memory in Japan' in Hui, T.Y., van Bremen, J. and Ben-Ari, E. (eds) (2005) *Perspectives on Social Memory in Japan*, Folkestone: Global Oriental.
van Wolferen, K. (1989) *The Enigma of Japanese Power: People and Politics in a Stateless Nation*, London: Macmillan.
Video Research (1985) *Video Research Digest*, August.
— *(2009)* www.videor.co.jp/data/ratedata/backnum/2009/vol32.htm, accessed 24 August 2009.
Vinitzky-Serousii, V. (2005) 'Social Memory and Commemoration: Some "After Fact" Thoughts' in Hui, T.Y., van Bremen, J. and Ben-Ari, E. (eds) (2005) *Perspectives on Social Memory in Japan*, Folkestone: Global Oriental.
Waddington, K. (2009) 'Not for ourselves, but for others': The rhetoric of charity and social display, in Liedtke, R. and Weber, K. (eds), *Philanthropie und Religion in den europäischen Zivilgesellschaften: Entwicklungen im 19. und 20. Jahrhundert*, Paderborn: Schöningh.
Walter, T. (1993) 'War Grave Pilgrimage' in Reader, I. and Walter, T. (eds) (1993) *Pilgrimage in Popular Culture*, Basingstoke: Macmillan.
— (2009) 'Dark Tourism: Mediating between the dead and the living' in Sharpley, R. and Stone, P.R. (eds) (2009) *The Darker Side of Travel*, Bristol: Channel View.
Weiner, M. and Chapman, D. (2009) 'Zainichi Koreans in history and memory' in Weiner, M. (ed.) (2009) *Japan's Minorities*, London: Routledge.
Weir, A. (2000) *The Tombstone Imperative: The Truth about Air Safety*, London: Simon & Schuster.
West Sellers, R. and Walter, T. (1993) 'From Custer to Kent State: Heroes, Martyrs and the Evolution of Popular Shrines in the USA' in Reader, I. and Walter, T. (eds) (1993) *Pilgrimage in Popular Culture, Basingstoke: Macmillan.*
Wikipedia (2010) en.wikipedia.org/wiki/Kyu_Sakamoto#Death, accessed 1 December 2010.
Wiegmann, D.A. and Shappell, S.A. (2003) *A Human Error Approach to Aviation Accident Analysis: The Human Factors Analysis and Classification System*, Aldershot: Ashgate.
Winslow, J. (2002) *Mayday*, Fyshwick: Aerospace Publications.
Wisner, B., Blaikie, P., Cannon, T. and Davis, I. (2008) *At Risk*, Second edition, London: Routledge.
Yamamoto, Y. (1999a) *Tsuiraku no Haikei – Nikkōki-wa Naze Ochitaka*, vol. 1, Tōkyō: Kōdansha.
— (1999b) *Tsuiraku no Haikei – Nikkōki-wa Naze Ochitaka*, vol. 2, Tōkyō: Kōdansha.

— (2001) *Nihon Kōkū Jiko Shori Tantō*, Tōkyō: Kōdansha.
Yamazaki, A. (2008a) *Naze Bōingu-wa Ikinokottanoka*, Tōkyō: Ei Shuppansha.
— (2008b) *Naze Eabasu-wa Katanainoka*, Tōkyō: Ei Shuppansha.
Yamazaki, T. (2001) *Shizumanu Taiyō* (5 volumes), Tōkyō: Shinchōsha.
Yanagida, K. (1988) *Shikaku – Kyodai Jiko no Gemba*, Tōkyō: Shinchōsha.
— (1991) *Jijitsu no Kanagekata*, Tōkyō: Shinchōsha.
— (1992) *Gijutsu Shakai no Kage*, Tōkyō: Bungei Shunjū.
— (2009) *Ikinaosu Chikara*, Tōkyō: Shinchōsha.
Yokoyama, H. (2006) *Kuraimāzu Hai*, Tōkyō: Bungei Shunjū.
Yomiuri Shimbun (2010) adv.yomiuri.co.jp/yomiuri/busu/busu01.html, accessed 1 December 2010.
Yoneda, K. (2005) *Osutaka-yama no Nazo-o Ou – Nikkō 123bin Jiko 20 Nen*, Tōkyō: Takarajimasha.
Yoshida, K. (2006) *Yumiko-e*, Tōkyō: Fusōsha.
Yoshida, T. (2006) *Ihin Seiriya-wa Mita!*, Tōkyō: Fusōsha.
Yoshihara, K. (1985a) *Ima Hikōki-wa Anzenka*, Tōkyō: Sanseidō.
— (1985b) *Jambo Tsuiraku – unable to control! (1985)*, Tōkyō: Ningen-no Kagakushinsha.
— (2003) *Waza to Chie no Sōshitsu*, Tōkyō: Rukku.
— (2004) *Jambo Tsuiraku*, Tōkyō: Ningen-no Kagakushinsha.
Yoshioka, S. (1986) *Tsuiraku no Natsu – Nikkō 123bin Jiko Zenkiroku*, Tōkyō: Shinchōsha.
Yoshizaki, T. (2005) *1985nen*, Tōkyō: Shinchōsha.
Young, J.E. (1993) *The Texture of Memory: Holocaust Memorials and Meaning*, New Haven: Yale University Press.

Index